ANALYSIS OF POLYMER SYSTEMS

ANALYSIS OF POLYMER SYSTEMS

Edited by

L. S. BARK

Department of Chemistry and Applied Chemistry,
University of Salford, Salford, UK

and

N. S. ALLEN

Department of Chemistry, John Dalton Faculty of Technology,
Manchester Polytechnic, Manchester, UK

APPLIED SCIENCE PUBLISHERS LTD
LONDON

APPLIED SCIENCE PUBLISHERS LTD
Ripple Road, Barking, Essex, England

British Library Cataloguing in Publication Data

Analysis of polymer systems.
1. Polymers and Polymerization—Analysis
I. Bark, L. S.   II. Allen, N. S.
547.8′4   QD139.P6

ISBN 0-85334-122-2

WITH 18 TABLES AND 102 ILLUSTRATIONS

Printed in Great Britain by Galliard (Printers) Ltd, Great Yarmouth

FOREWORD

In common with all modern industrial processes and derived materials, it is essential in the case of synthetic polymers to have an accurate knowledge of the materials being used and produced at all stages of the manufacturing process. For polymers, which now help to form a major part of all our lives, the end-product can only be acceptable providing that the intermediate products also are. The uses and applications of new polymeric materials can only be placed on a scientific basis if a knowledge of their composition, in all aspects, accompanies the pragmatism of field tests.

However, as scientists widen the scope of the analytical sciences (and of analytical chemistry in particular) it becomes necessary to examine the methods available for the analysis of this type of material and to assess or even re-assess the rôle of any particular method.

Modern industrial processes often require a speed of attainment of knowledge that cannot wait upon chemical reaction and hence physical comparative methods may be of great importance when dealing with 'pure' materials. Such appropriate methods are dealt with in detail in several chapters.

Some methods have not yet been fully exploited and some authors have indicated potential uses for selected techniques. It is hoped that fellow-workers will see the potentialities for their own problems.

However, in spite of all these new methods, it is still essential for an unknown material to have some 'narrowing down' of the possibilities. The first chapter, dealing with 'preliminary tests', may seem to be chrono-logically out of line; we justify this by practice. It has been found that it is still economically viable to have a 'general analyst' for unknown samples. The 'general analyst' can indicate what techniques are most likely to produce the required results in the required time.

It is our opinion that few analyses are done without some prior knowledge of the nature or uses of the polymer. (Exceptions are, of course, in the fields of forensic science and pathological chemistry.) In general, the polymer chemist/scientist must establish a dialogue with his counterparts in analytical sciences. This book is intended to help him achieve this aim.

L. S. BARK

PREFACE

The analysis of polymer systems continues to be a rapidly developing field of research particularly as many new applications become available with both new and established techniques that are capable of handling many of the intractable polymer systems. This book deals with such important applications for a number of well established and new techniques.

The book contains a total of eight chapters each dealing with the application of a particular technique for the analysis, characterisation (of additives, molecular weight and structure) and identification of polymer systems.

Chapter 1 written by Professor Bark (UK) covers one of the most important aspects of polymer analysis and deals with the numerous preliminary tests such as elemental analysis and polymer identification. The microstructural analysis of polymer systems using high resolution NMR spectroscopy is an area of very rapid development and in Chapter 2 Dr Ebdon (UK) discusses some of the more important established developments. Infrared and Raman spectroscopy are well established tools in polymer analysis and in Chapter 3 some of the important applications are dealt with by Dr Maddams (UK). Luminescence spectroscopy is a relatively new technique in the field of polymer analysis and is finding widespread application as a means of identification, quantitative analysis of additives and for probing the conformation of polymer chains. This technique is covered by Dr Allen (UK) in Chapter 4. Probably one of the most difficult techniques to be applied to the study of polymers is mass spectrometry. However, through the use of some carefully designed modifications the technique has found widespread use as a means of identifying and probing polymeric structures. The technique has also found widespread use for investigating the mechanisms of polymer degradation; all these aspects are dealt with in Chapter 5 by Drs Foti and Montaudo

(Italy). Thermal methods of analysis are now almost routine in most polymer laboratories and in Chapter 6 Dr Hay (UK) discusses some of the more important applications of the techniques involved. Some important applications in the measurement of polymer molecular weights using ebullioscopic methods are discussed by Dr Davison (UK) in Chapter 7 and finally one vitally important technique for both molecular weight determinations and separations is gel permeation chromatography dealt with by Dr Cooper (USA) in Chapter 8.

The book is not meant to be a complete text on polymer analysis but is meant rather to create an awareness in polymer analysts of the more important applications and developments that have been made in the field. The book will be of wide interest to all polymer scientists and technologists, analytical chemists in the polymer industry, government laboratories and academic institutions and colleges. The book, we hope, will also be of value for tutors in the teaching of polymer analysis to undergraduates and postgraduates. Finally, we welcome any comments that readers may have.

L. S. BARK
and
N. S. ALLEN

CONTENTS

LIST OF CONTRIBUTORS

NORMAN S. ALLEN

Department of Chemistry, Manchester Polytechnic, John Dalton Faculty of Technology, Chester Street, Manchester M1 5GD, UK.

L. S. BARK

Department of Chemistry and Applied Chemistry, University of Salford, Salford M5 4WT, UK.

ANTHONY R. COOPER

Chemistry Department, Lockheed Missiles and Space Company, Inc., 3251 Hanover Street, Building 204, Palo Alto, California 94304, USA.

G. DAVISON

34 Beechfields, Doctors Lane, Eccleston, Chorley, Lancashire PR7 5RE, UK.

J. R. EBDON

Department of Chemistry, University of Lancaster, Bailrigg, Lancaster LA1 4YA, UK.

SALVATORE FOTI

Istituto Dipartimentale di Chimica e Chimica Industriale, Università di Catania, Viale A. Doria 8, 95125 Catania, Italy.

JAMES N. HAY

Department of Chemistry, University of Birmingham, Birmingham
B15 2TT, UK.

W. F. MADDAMS

Analytical Services, British Petroleum Research Centre, Chertsey
Road, Sunbury-on-Thames, Middlesex TW16 7LN, UK.

GIORGIO MONTAUDO

Istituto Dipartimentale di Chimica e Chimica Industriale, Università
di Catania, Viale A. Doria 8, 95125 Catania, Italy.

Chapter 1

PRELIMINARY TESTS

L. S. BARK

*Department of Chemistry and Applied Chemistry,
University of Salford, UK*

1.1. INTRODUCTION

Although the analysis of polymeric materials covers several important aspects there are, as always for the analytical chemist when confronted with a problem, a number of fundamental questions to be answered. These questions, of course, must be answered before detailed quantitative information is obtained. There is no purpose in attempting detailed esoteric investigations unless it has been previously ascertained what type of polymer or what mixture of polymers is present and what other materials, such as various fillers, are being used.

It is important to remember that the purpose of any analysis is to provide an answer to a problem which may be as simple as providing a decision on the suitability or otherwise of a particular polymer for a chosen purpose.

In a number of cases the industrial chemist may not know what type of polymer is being used, and under these circumstances the analytical chemist must supply this information.

There are, of course, a number of fundamental questions which all analytical chemists must ask when supplied with a sample for analysis, for example, 'Is the sample a "representative sample"?'. Inherent in this question are others, including 'Are all the constituents present in this sample in the same composition, proportions and chemical environment as in the original material?', or 'Did the method of sampling result in chemical changes in the material?'.

The methods used for sampling must be carefully considered and controlled. If the polymeric material is a liquid and is to be used in bulk then homogenisation before sampling is necessary and is generally simply obtained by rapid stirring, either by a mechanical means or by ultrasonic agitation.

For solid samples the material may have to be reduced to a powder. Occasionally this may be achieved by crushing the material in a mortar. If necessary, the sample can be cooled in a suitable freezing mixture, such as liquid air, to render it brittle. If this is not possible then filing, drilling, cutting in a mill, or turning on a lathe may be used to obtain a finely divided specimen. It must be remembered that the material may become hot during these operations and it is necessary to proceed with caution to avoid changes in chemical composition; this is important for polymers. A visual examination using a low powered microscope or even a hand lens is useful to ascertain whether or not melting has occurred during the sampling and enable the procedures to be modified so as to ensure no thermal decomposition has taken place.

Having obtained a representative sample then a systematic scheme of testing is generally valuable in enabling a qualitative identification of the polymer to be obtained with minimal expenditure of effort, equipment, time and hence cost. Industrial polymers are not generally single substances. Indeed, the majority of commercially used polymers contain, in addition to low molecular mass fractions including monomer and catalyst remnants, the dyes, pigments, fillers, plasticisers, solvents which are used to make the product commercially viable. In many cases there are stabilisers to prevent cross-linking or unzipping of polymers during exposure to u.v. radiation and/or thermal processing. An additional complication is that variations in the degree of crystallinity, copolymerisation, etc., cause variations in both the chemical and the physical properties of the polymer, and hence sample tests based on parameters such as melting point, glass transitional points, density, etc., must be used with great caution.

With materials of such complexity it is essential to separate the various components and then systematically investigate each of them as appropriate. It is also necessary to select, whenever possible, separation methods which do not chemically alter the polymer and hence solvents which react with the polymer should be avoided whenever possible.

Generally it is advisable to remove the plasticiser as the first step in the separation procedure. The plasticiser is extracted from the polymer sample using a soxhlet extraction system with a suitable solvent such as diethyl ether or a low boiling point petroleum ether.

The inorganic fillers and the polymer are then separated by subsequent dissolution of the material. It is often necessary to investigate the use of various solvents in order to obtain satisfactory dissolution of the polymer and this should be done in the usual way by gently warming a small amount of the mixture of the fillers and the polymer with a few millilitres of the solvent, removing the residue from the mixture by filtration, and the solvent from the filtrate by evaporation. Visual examination of the residue will confirm the efficacy of the chosen solvent. When a suitable solvent has been found and the polymer has been separated from the fillers it may be recovered, either by removal of the solvent by evaporation under reduced pressure or by precipitation on the addition of a 'non-solvent' liquid. (The latter method can sometimes result in the formation of an 'oily' residue and is not preferred by the present author.)

The separated filler can, of course, be investigated by the usual chemical or spectrographic procedures. (If confirmation is required it may be necessary to destroy the organic matter in a separate procedure (see later).) Thermosetting insoluble plastics present a somewhat different problem since fillers cannot normally be separated by extraction procedures. In this case the plastic will have to be modified by chemical reaction, for example, either esterification or occasionally hydrolysis may be useful. If the filler is organic in nature then it is not always possible to separate it from the polymer since many polymers and organic fillers are chemically related.

In such cases it is usual to attempt the purification of the polymer by repeated dissolution in and precipitation from selected solvents. Many processes, evolved by trial and error, have been proposed. A well established procedure[1] still serves as an illustrative example for polystyrene. Dissolve the sample in toluene at the reflux, separate any residue by filtration or decantation, and then precipitate the polystyrene by dropwise addition of isopropanol. Remove the precipitated polymer, wash it with warm isopropanol and dissolve it in ethyl methyl ketone, reflux the mixture, separate any residue, allow the solution to cool below 50°C and then reprecipitate the polystyrene by the dropwise addition of methanol. Wash the precipitate with methanol and repeat the whole procedure.

By such 'fractional precipitation' most of the organic fillers and low molecular mass contaminants are removed. Care must be taken when drying the polymer, hence the use of a low volatile solvent in the last wash. The material should finally be dried under vacuum.

1.2. SYSTEMATIC ANALYSIS OF THE ISOLATED POLYMER

In the analysis of any polymer it is essential to know what chemical functionalities are present, since a knowledge of these not only helps to classify the material but also helps in the prediction of properties and in the forming of opinions concerning the suitability of the product for its proposed commercial usage.

Although many techniques (described in later chapters) are available for the precise definition of a polymer, it is important that the user of such techniques has some pre-knowledge of what to look for, what to measure and, perhaps most important, what chemical functionalities are not present. Most techniques described later operate on the basis of the interpretation of an electrically generated signal which may be selectively 'generated' by a particular functionality. For example, the —NH group in polyamides absorbs infrared radiation in the region $3050\text{--}3470\,\text{cm}^{-1}$, whilst the —OH group absorbs in the region $3360\text{--}3760\,\text{cm}^{-1}$; the regions of maximum absorption may be modified by the particular environment in the sample to hand. Similarly, polyamides and polyketones exhibit strong infrared absorption bands in the region $1600\,\text{cm}^{-1}$. It is, therefore, of use to know whether or not the polymer contains nitrogen and, if possible, in what chemical (and hence infrared) functionality before drawing conclusions from a single electrical signal.

1.3. QUALITATIVE IDENTIFICATION OF THE ELEMENTAL NATURE OF THE POLYMER

It must be emphasised that simple qualitative tests to ascertain the nature of the polymer are not specific and it is necessary to use all schemes with caution. It is essential, for any scheme, to have a qualitative knowledge of the elemental composition of the polymer in order to decrease the number of probable interpretations of the results of any single test.

The elements which may be contained in a polymer are, in decreasing order of probability: carbon, hydrogen, oxygen, nitrogen, chlorine, fluorine, sulphur, silicon, phosphorus, titanium and boron. (Occasionally bromine is encountered.) It is, therefore, essential to establish which of these elements are present. All the elements which are present in any organic functionality are converted into an inorganic compound whose presence is then easily detected by sensitive and unambiguous tests.

There are various methods available for this organic–inorganic conversion, and care must be taken in the choice of method. One of the main problems is that of sensitivity. The polymeric material separated from the sample may often be a mixture of polymers. Some components of the mixture may be in proportions which are industrially (and hence analyti-cally) significant but which do not allow for easy chemical detection, especially when the usual weight taken for all such tests is about 20 mg.

For example, if the mixture contains less than 5 % by weight of a polymer based on acrylonitrile or a polyamide, in the presence of large amounts of a vinyl chloride or chlorinated polyethylene polymer, then the amount of nitrogen is very small and may be easily missed in the presence of such large amounts of chlorine.

Whilst two of the commonly used tests for detection of the elements are Lassaigne's test (sodium fusion) and fusion with a mixture of zinc dust and sodium carbonate, both are relatively insensitive and certainly do not give complete conversion to the inorganic species.

Several workers recommend complete oxidation either by fusion with sodium peroxide in a Parr bomb or by combustion in an oxygen flask (for a detailed account of the use of the latter see ref. 2).

Essentially in this method the polymer is wrapped in ashless filter paper which is ignited and burnt in a closed glass flask filled with purified oxygen in the presence of a solution of sodium hydroxide which is used to absorb the acidic compounds produced. (The ignition may be caused by an electric discharge.[3]) The procedure produces the sodium salts of nitric acid (from N), hydrofluoric, hydrochloric, hydrobromic acids (from F, Cl and Br), sulphurous and sulphuric acids (from S), and phosphoric acid (from P). The solution is then divided into aliquots for the detection of the individual elements by standard tests.

Chloride and bromide may be detected by the precipitation reaction with silver ions. Haslam et al.[2] recommends the use of a colorimetric reaction in which thiocyanate is displaced from mercury (II) thiocyanate in the presence of iron (III) and the appearance of a red colour indicates the presence of chloride or bromide. Whilst this has the advantage that it can be made semi-quantitative, it does not differentiate between the two halogens.

Fluorine is best determined by the use of the cerium alizarin complexone reaction devised by Belcher et al.[4]

Nitrogen is detected in the form of nitrate by one of several tests, of which the simple 'brown ring test' is probably the most rapid. The reaction of nitrate with a polyhydric phenol (resorcinol) in the presence of iron (II) to give a green colour may also be used.

Sulphur, which is converted to sulphite, and sulphate are best detected by first ensuring complete conversion to sulphate by boiling the aliquot from the combustion flask with hydrogen peroxide, before acidification with hydrochloric acid and precipitation of any sulphate with barium chloride solution.

Phosphorus may be detected by the precipitation as an ammonium phosphomolybdate complex by standard methods. This precipitation is insensitive and a better method is to reduce the complex *in situ* with a suitable reducing agent (tin (II) or ascorbic acid) when a blue colour (from the formation of a 'molybdenum blue') indicates the presence of phosphorus in the original sample. Silicon (from silicones) interferes with this test.

It must be emphasised that it is essential in all these tests to carry out a blank determination, viz., combust only the filter paper and then proceed with aliquots from the resultant solution as previously indicated.

This procedure cannot be used for the detection of either boron or silicon since there are inevitably traces of these present in the test solution as a result of dissolution of some of the borosilicate of the 'Pyrex' or equivalent glass used for the combustion vessel. Although it has been suggested that it is possible to have an indication of the presence of either by comparing colours produced from the test solution and the blank solution in a series of tests, or even ensuring that boron is absent by replacing all borosilicate parts of the apparatus with fused silica, there is no completely satisfactory and simple routine method for the detection of silicon.

Since both phosphorus and silicon give molybdate complexes it is necessary to remove the phosphorus from the test solutions before testing for silicon. This is best done by using the solution from a Parr bomb.

The above tests indicate whether or not the polymer contains elements other than carbon, hydrogen and oxygen; it is however useful to be able to differentiate between hydrocarbons and organic compounds containing oxygen. A sensitive test is that devised by Feigl,[5] in which a drop of a saturated solution of the polymer in chloroform is stirred with a thin glass rod covered with the iron (III) thiocyanato complex (prepared by mixing aqueous potassium thiocyanate solution with aqueous iron (III) chloride solution and extracting the complex with ether). The glass rod is dipped into the ether extract and allowed to dry. A positive test is obtained if the drop becomes light to dark red.

Similarly, if the chloroform solution is added to a finely ground mixture of equal weights of potassium thiocyanate and iron (III) alum, a reddish-purple colour in the organic layer indicates the presence of oxygen in the compound.

Esters and alcohols give a positive test. Saturated and unsaturated hydrocarbons and their halogenated derivatives do not give a positive response. Whilst a positive result is always reliable, some oxygenation compounds fail to give a positive test and the test is only reliable if sulphur and nitrogen are absent.

1.4. USE OF THE RESULTS FROM ELEMENTAL ANALYSIS

Since the formulation of polymer mixes is ever-changing, and the type of polymer encountered continues to vary as the cost of production varies, it is not possible to give a comprehensive list of polymers classified according to their elemental composition. Table 1.1 indicates the most commonly occurring polymers in order of elemental composition.

1.5. THERMAL DECOMPOSITION OF POLYMERS

In preliminary and non-quantitative investigations there are tests and procedures for two types of thermal decomposition, these involve (a) burning, and (b) pyrolysis of the polymer. Each procedure both supplements and complements the other and the results must be considered together in order to obtain the maximum information. Of all the tests used in characterising polymers, these thermal decomposition tests require the analyst to have a good sense of smell, a good memory for smells, *and previously obtained experience on known polymeric material.* It is necessary to have both patience and persistence to obtain the maximum amount of information since many individual phenomena may be observed in any one test.

Since these tests are the most subjective they are often a source of controversy, but in the hands of experienced analysts they are capable of giving a great deal of qualitative information in a very short time.

1.5.1. Burning Characteristics

A small amount of the sample (0·1–0·2 g) is held on a spatula in the edge of a small non-luminous flame from a micro bunsen burner. If the sample does not ignite immediately it should be held in the flame for a few seconds and then withdrawn.

The ease of ignition, any odours given off, the self-extinguishing characteristics, any colour changes, whether or not the sample melts sharply, with or without decomposition, chars or burns, must be observed. The colour of the flame and the formation and nature of any fumes produced must be examined. Finally, the sample should be ignited to determine the nature of any ash.

Since no valid conclusions can be obtained from smelling the polymer

TABLE 1.1
The Most Commonly Occurring Polymers

Elements (*other than C and H*)	*Polymers*
None	Polymerised hydrocarbons (e.g. polyethylene, poly-propylene, polybutadienes, polystyrene, poly(methyl styrene), poly(divinyl benzene), polybutadiene-styrene)
O	Carbohydrates (e.g. celluloses, starches); Phenol-formaldehyde resins, phenol-ether resins, phenol-furfural resins, phthalic acid esters and related compounds; Alkyds; Polyvinyl esters; Polyallyl esters; Polyacrylic and related esters; Polycarbonates; Aliphatic polyethers; Epoxy resins; Aldehyde resins; Poly(vinyl alcohol)
N (O)	Cellulose nitrate; Polyacrylonitrile and copolymers; Polyacrylamide; Nylons and other polyamides; Poly-ethylene imines; Urea-formaldehyde resins; Melamine-formaldehyde resins; Polyurethane; Polyester–urethane
Cl (O)	Chlorinated polyhydrocarbons, poly(vinyl chloride), poly(vinylidene chloride), poly(vinyl chloride acetate), polychloroacrylates, chlorinated alkyds
F (O)	Polytetrafluoroethylene; Poly(vinyl fluoride)
F, Cl	Polychlorotrifluoroethylene
S (O)	Alkyl polysulphides; Polysulphones; Thiophenol condensates
N, S	Thiourea-formaldehyde resins
Cl, S	Thioplasts; Vulcanised neoprenes
N, S, P	Casein; Protein
Other elements: Si	Silicones
Ti	Polytitanic esters
B	Polyboric esters
N, P, Cl	Polyphosphonitrilic halides

This list is *not* intended to be comprehensive.

whilst it is burning to detect any odours produced, any flames produced are extinguished and the odour of the rising vapour observed. There may be successive stages of decomposition, therefore the spatula should be returned to the edge of the flame and removed at frequent intervals to ascertain whether or not burning is occurring and if there are any changes in the fumes evolved. It must also be remembered that many of the vapours and fumes are unpleasant and acrid, and too much vapour may deaden the sense of smell.

The observations should therefore be made in the following order:

(i)   Ease or difficulty of ignition.
(ii)   Whether or not the flame is self-extinguishing.
(iii) The character of the flame; this includes the colours, whether the flame is soot-producing.
(iv)   The physical behaviour of the sample; that is, whether it swells, softens, melts, chars readily, leaves a residue.
(v)   The odour of any fumes evolved.

There are various systems used for the next stage of the examination and it is probable that each analyst devises his own, depending to a large extent upon the general nature of the polymers presented to him. Those systems based on the elemental composition of the polymer are generally the most convenient to use.

It must be emphasised that the sense of smell is extremely subjective and in many cases the only way to describe the smell is 'characteristic'. The identification of odours is a matter of experience and of practice, and methods of identification based solely or even mainly on recognition or identification of an odour are far too subjective to enable any satisfactory scientific conclusion to be obtained.

Known specimens should be decomposed and compared with the unknown. Mixtures tend to give confusing results.

1.6. CLASSIFICATION ACCORDING TO ELEMENTAL COMPOSITION AND BEHAVIOUR ON HEATING

1.6.1. Polymers Containing Carbon and Hydrogen Only, or Carbon, Hydrogen and Oxygen Only (Table 1.2)

The most commonly encountered are polyethylenes, polypropylenes, and α-olefin polymers, starches and polymers, methacrylates and related compounds and epoxy resins.

TABLE 1.2
Polymers Containing Carbon and Hydrogen Only, or Carbon, Hydrogen and Oxygen Only

Ease of ignition	Flame characteristics	Physical behaviour of material	Characteristic odours	Probable polymer type
Very difficult	(i) Yellow flame (ii) Self-extinguishing	Retains general shape, even though some swelling and cracking	Formaldehyde and phenol (second)	Phenol-formaldehyde
Difficult	(i) Yellow–orange with black smoke (ii) Continues to burn	Softens, melts, becomes relatively mobile; leaves hard misshapen beads	Sweet, slightly aromatic	Poly(ethylene terephthalate)
Moderate to difficult	(i) Yellow; bluish base (ii) Self-extinguishing	Swells, cracks, flame spurts	Burning wood	Lignocellulose
	(i) Pale yellow with much soot (ii) Slowly self-extinguishes	Melts and then chars	Not characteristic	Benzyl cellulose
	(i) Yellow; grey–black smoke (ii) Slowly self-extinguishes	Swells, softens, decomposes via light-brown mass	Pungent	Poly(vinyl alcohol)
Moderate	(i) Yellow flame, some soot (ii) Slowly self-extinguishes	Swells, cracks, chars	Phenolic	Phenolic laminates
	(i) Yellow flame with blue base. Very little smoke (ii) Continues burning	Melts, flows easily	Burning paraffin	Polyethylene, polypropylene
	(i) Yellow flame with grey smoke (ii) Continues burning	Does not melt easily; softens, burns steadily	Acrid or slightly bituminous	Polyester
	(i) Yellow–orange with blue base (ii) Continues burning	Melts, flows easily	Butyrates (rancid butter)	Poly(vinyl butyral)
	(i) Luminous flame, very little smoke (ii) Continues burning	Melts, decomposes with white vapour evolved	Acrid (acrolein)	Some alkyd resins

Moderate–Readily	(i) Yellow–grey smoke (ii) Continues burning	Melts—sometimes spurts of flame	Not characteristic[a]	Epoxides
Readily	(i) Yellow-orange flame with much soot (ii) Continues burning	Softens, but does not readily flow	Sweet and characteristic[a]	Polystyrene
	(i) Dark yellow flame with moderate amounts of smoke (ii) Continues burning	Softens, but does not readily flow	Acetic acid vapours	Poly(methyl styrene) Poly(vinyl acetate)
	(i) Luminous flame (ii) Continues burning	Melts readily. Decomposes with visible fumes but with little charring	Pungent and acidic	Polyacrylate
	(i) Yellow flame, some smoke (ii) Continues burning	Softens; does not flow easily. Very little charring	Sweet, fruity[a]	Poly(methyl methacrylate)
	(i) Yellow–green flame (ii) Continues burning	Melts readily and chars	Sweet smell	Methyl cellulose Ethyl cellulose
	(i) Yellow flame with blue base (ii) Continues burning	Melts readily, burns quickly. Charred bead (non-regular) left	Acetic acid	Cellulose acetates
	(i) Yellow flame with some smoke (ii) Continues burning	Burns very fast. Can leave very brittle charred mass	Burning paper	Cellophane
	(i) Luminous flame (ii) Continues burning	Melts and flows	Rubber-like	Poly-isobutylenes
Very readily	(i) Yellow flame (ii) Continues burning	Burns rapidly to dry ash	Burning paper	Cellulose

[a] Known samples should be tested if these materials are suspected.

1.6.2. Polymers Containing Carbon, Hydrogen (Oxygen) and Nitrogen (Table 1.3)

Probably the most commonly encountered polymers of this class are the nylons, the polyurethanes and isocyanates. These have replaced, to a large extent, the urea-formaldehyde and melamine-formaldehyde resins as the main bulk use for nitrogen-containing resins. Polyacrylonitriles and polyimides are now being used in greater amounts. It is still possible to encounter casein/casein-based resins as well as cellulose nitrate resins, although the latter are being discontinued for many uses because of their very high flammability.

1.6.3. Polymers Containing Carbon, Hydrogen (Oxygen), Nitrogen and Chlorine (Table 1.4)

The main polymers containing both nitrogen and chlorine are based on poly(vinyl chloride acrylonitrile). Although these are not strictly compounds, but physical mixtures, they are difficult to separate into individual components and for practical purposes should be regarded as compounds.

1.6.4. Polymers Containing Carbon, Hydrogen (Oxygen) and Halogens (Table 1.5)

For most practical purposes it is probable that if the polymer contains only fluorine (of the halogens) then the polymer will be either polytetrafluoroethylene or poly(vinyl fluoride). Some fluorocarbon polymers and copolymers are found (e.g. hexafluoroprene/vinylidene fluoride (Viton A), hexafluoroproprene/tetrafluoroethylene (Teflon FEP)) but these are indistinguishable from each other by simple methods.

Polymers containing only chlorine (of the halogens) are probably poly(vinyl chloride)/related compounds or poly(vinylidene chloride) and related compounds.

It is likely that polymers containing both fluorine and chlorine are either polytrifluorochloroethylene or compounds such as trifluorochloroethylene/vinylidene chloride (KEL F elastomer), and these are often found in small amounts in paints and similar materials.

1.6.5. Polymers Containing Other Elements

1.6.5.1. Sulphur

There are only a few types of polymer which contain only carbon, hydrogen, (oxygen) and sulphur, and these are generally rubber-type polymers which have been vulcanised, or polysulphones and polysulphides. In the cases where both nitrogen and sulphur have been detected then the

TABLE 1.3
Polymers Containing Carbon, Hydrogen (Oxygen) and Nitrogen

Ease of ignition	Flame characteristics	Physical behaviour of material	Characteristic odours	Probable polymer type
Very difficult	(i) Light yellow flame (ii) Self-extinguishing	Retains shape, but may swell and crack	Formaldehyde followed by amine-like smell	Melamine-formaldehyde
	(i) Yellow–orange flame (ii) Self-extinguishing	Retains shape, but may swell and crack	Formaldehyde only	Urea-formaldehyde
Difficult	(i) Yellow flame (ii) Self-extinguishing	Melts to give droplets, froths	Burning vegetables	Polyamides (Nylons)
Moderate	(i) Yellow flame. Grey smoke (ii) Self-extinguishing	Swells, chars, froths; crisp char left	Burning hair	Protein
Easily	(i) Yellow–orange (ii) Continues burning	Melts and flows easily	Cyanide and burning wood	Isocyanate
	(i) Yellow–orange (ii) Continues burning	Melts, some charring	Characteristic[a]	Polyimides
Very easily	(i) Burns very rapidly with white flame (ii) Continues burning	Curls up, slight charring, some melting	Acrid—very pungent	Cellulose nitrate

[a] Known samples should be tested if these materials are suspected.

TABLE 1.4

Polymers Containing Carbon, Hydrogen (Oxygen), Nitrogen and Chlorine

Ease of ignition	Flame characteristics	Physical behaviour of material	Characteristic odours	Probable polymer type
Difficult	(i) Yellow with green edge (ii) Self-extinguishing	Shrinks noticeably. Softens and melts, does not flow, and leaves hard bead	Characteristic[a]	Poly(vinyl chlorine acrylonitrile)

[a] Known samples should be tested if these materials are suspected.

TABLE 1.5

Polymers Containing Carbon, Hydrogen (Oxygen) and Halogen

Ease of ignition	Flame characteristics	Physical behaviour of material	Characteristic odours	Probable polymer type
Very difficult	(i) Yellow–orange, edged with green (ii) Self-extinguishing	Black, hard residue	Chlorine-type smell	Poly(vinylidene chloride)
	(i) Yellow–grey smoke (ii) Self-extinguishing	Softens, does not flow	Very little, but acrid	Polytetrafluoroethylene
Difficult	(i) Yellow–green edge (ii) Self-extinguishing	Darkens, becomes black and hard	Acrid–chlorine	Poly(vinyl chloride)
	(i) Yellow–green edge, sooty flame (ii) Self-extinguishing	Softens, does not char	Acrid–chlorine	Poly(vinyl chloride)/(acetate)

TABLE 1.6
Pyrolysis of Polymers

	Observations (odour)	Tests	Conclusions
(i)	Smell of burnt paper	Vapours burn with characteristic flame—yellow with little smoke	Cellulose
(ii)	(i) + smell of acetic acid		Cellulose acetate
(iii)	(i) + smell of butyrates		Cellulose butyrate
(iv)	Smell of acetic acid	Fumes acidic to BDH Universal Indicator Paper	Acetate polymers
(v)	Cyanide		Polyacrylonitrile
(vi)	Burning vegetation	Test vapour with damp Universal Indicator Paper. Alkaline vapours	Polyamides (nylon), polyurethanes, protein-type resins
(vii)	Smell of formaldehyde		Urea-formaldehyde
(viii)	(vii) + amine-like smell		Melamine-formaldehyde
(ix)	(vii) + phenolic smell		Phenol-formaldehyde
(x)	Sweetish odour		Methacrylate and related compounds
(xi)	Acrid odour	Allow vapours to impinge on a glass rod moistened with silver nitrate solution. Cloudy precipitate	Poly(vinyl chloride) and its copolymers. Natural and synthetic chlorinated rubbers, chloroprene and chlorinated polyolefins
(xii)	(xi) + heavy black ash	Allow vapours to impinge on a glass rod moistened with silver nitrate solution. Cloudy precipitate	Probably poly(vinylidene chloride)
(xiii)	Sulphurous	Allow vapours to impinge on a glass rod moistened with barium chloride solution. Cloudy precipitate	Sulphur-containing, rubbers, etc.

N.B. Other indications are usually not informative except when used by very experienced operators.

polymer is probably a thiourea-formaldehyde resin. The only commonly occurring polymers containing sulphur, nitrogen and phosphorus are those based on casein, and tests for proteins should then be done.

1.6.5.2. Phosphorus and Silicon
Polymers containing phosphorus only and silicon only, should be examined by physical methods since they are probably phosphorus esters or silicones, and tests involving heating, burning and pyrolysis will give no further information other than this.

1.6.6. Pyrolysis of Polymers
The use of automated pyrolysis systems having strictly controlled physical parameters such as temperature, removal of gaseous products from the vicinity of the heated polymer, followed by gas chromatographic separation of the products and ultimately identification by mass spectroscopy has much to recommend it and is very highly regarded. It must, however, be emphasised that strict control of all conditions is essential if absolute identification is required, since the products formed will depend not only on the pyrolysis temperature but also on the speed of heating; for example, on slowly heating a relatively large mass of a carbon-containing polymer a large residue of carbonaceous char is produced from the recombination of products. The composition of the products will usually vary during the course of the pyrolysis and hence from general 'crude' laboratory conditions only general conclusions may be made.

The test is made by heating a small piece of the polymer in a small *dry* test tube (10 × 75 mm or *less*). At first a small flame is used, and then finally a fierce flame. The vapours given off are tested for acidity/alkalinity and odour. After testing for these parameters the vapours are ignited and the ease of ignition is noted. Finally the residue, if any, is examined visually (Table 1.6).

1.7.   IDENTIFICATION TESTS

As a result of the elemental analysis tests and the heating and pyrolysis tests it is possible to obtain a fairly substantive idea regarding the class of polymers into which to place unknown materials. There have been many systems proposed for the identification of polymers by chemical methods but these are considered to be outside the scope of this work. There are only a very few polymers for which specific chemical reactions are known and in

TABLE 1.7
Solubility Table

Polymer material	Soluble in	Insoluble in
Alkyd resin	Chlorinated hydrocarbons, lower alcohol, esters	Hydrocarbons
Amine-formaldehyde resin, cured	Benzylamine (60 °C), ammonia	
Cellulose ethers:		
Methyl cellulose	Water, dilute sodium hydroxide, methylene chloride, methanol	Acetone, ethanol
Ethyl cellulose	Methanol, methylene chloride, formic acid, acetic acid, pyridine	Aliphatic and aromatic hydrocarbons, water
Benzyl cellulose	Acetone, ethyl acetate, benzene, butanol	Aliphatic hydrocarbons, lower alcohol, water
Cellulose esters	Ketones, esters	Aliphatic hydrocarbons, water
Cellulose nitrate	Lower alcohols, acetic esters, ketones, ether/alcohols (3:1)	Ether, benzene, chlorinated hydrocarbons
Chlorine-containing polymers:		
Chlorinated rubber	Esters, ketones, carbon tetrachloride, tetrahydrofuran	Aliphatic hydrocarbons
Chloroprene rubber	Toluene, chlorinated hydrocarbons	Alcohols
Chlorinated polyethers	Cyclohexanone	Ethyl acetate, dimethyl formamide, toluene
Polychlorotrifluoroethylene	Hot fluorinated solvents	All common solvents
Poly(vinyl chloride)	Dimethyl formamide, tetrahydrofuran, cyclohexanone	Alcohols, butyl acetate, dioxan, hydrocarbons
Chlorinated poly(vinyl chloride)	Methylene chloride, cyclohexane, benzene, tetrachloroethylene	
Poly(vinylidene chloride)	Tetrahydrofuran, ketones, butyl acetate, dimethyl formamide (hot), chlorobenzene	Alcohols, hydrocarbons

TABLE 1.7—*contd.*

Polymer material	Soluble in	Insoluble in
Copolymers:		
Acrylonitrile/butadiene/styrene	Methylene chloride	Alcohols, water, aliphatic hydrocarbons
Styrene/butadiene	Ethyl acetate, benzene, methylene chloride	Alcohols, water
Vinyl chloride/vinyl acetate	Methylene chloride, cyclohexanone, tetrahydrofuran	Alcohols, hydrocarbons
Epoxy resins cured	Practically insoluble	
Fluorinated polymers:		
PTFE	Fluorocarbon oil (hot)	All solvents
Poly(vinyl fluoride)	Above 110 °C: cyclohexanone, propylene carbonate, dimethylsulphoxide, dimethyl formamide	
Poly(vinylidene fluoride)	Dimethylsulphoxide, dioxan	
Phenolic resin	Alcohol, ketones	Chlorinated hydrocarbons, aliphatic hydrocarbons
Poly(acrylic acid) derivatives:		
Polyacrylamide	Water	Alcohols, esters, hydrocarbons
Polyacrylonitrile	Dimethyl formamide, butyrolactone nitrophenol, dimethylsulphoxide, mineral acids	Alcohols, esters, ketones, formic acid, hydrocarbons
Poly(acrylic acid esters)	Aromatic hydrocarbons, esters, chlorinated hydrocarbons, acetone, tetrahydrofuran	Aliphatic hydrocarbons
Poly(methacrylic acid esters)	Aromatic hydrocarbons, dioxan, chlorinated hydrocarbons, esters, ketones	Ether, alcohols, aliphatic hydrocarbons
Polyamides	Phenols, formic acid, conc. mineral acids	Alcohols, esters, hydrocarbons
Polybutadiene	Aromatic hydrocarbons, cyclohexane, dibutyl ether	Alcohols, esters
Polycarbonates	Chlorinated hydrocarbons, dioxan, cyclohexanone	Alcohols, aliphatic hydrocarbons, water
Polyesters, unsaturated, uncured	Ketones, styrene, acrylic esters	Aliphatic hydrocarbons
Poly(ethylene terephthalate)	Cresol, conc. sulphuric acid, chlorophenol	
Polyethylene	Dichloroethylene, tetralin, hot hydrocarbons	Polar solvents, alcohols, esters

Polymer	Solvents	Non-solvents
Poly(ethylene glycol)	Chlorinated hydrocarbons, alcohols, water	Aliphatic hydrocarbons
Polyformaldehyde	Hot solvents, phenols, benzyl alcohol, dimethyl formamide	Alcohols, ketones, esters, hydrocarbons
Polyisoprene	Benzene	Alcohols, ketones, esters, hydrocarbons
Polypropylene	At high temperature: aromatic and chlorinated hydrocarbons, tetralin	Alcohols, esters, cyclohexanone
Polystyrene	Aromatic and chlorinated hydrocarbons, pyridine, ethyl acetate, methyl ethyl ketone, dioxan, tetralin	Alcohols, water, aliphatic hydrocarbons
Polyurethane	Tetrahydrofuran, pyridine, dimethyl formamide, formic acid, dimethylsulphoxide	Ether, alcohols, benzene, water, hydrogen chloride ($6N$)
Poly(vinyl acetal)	Esters, ketones, tetrahydrofuran	Methanol, aliphatic hydrocarbons
Poly(vinyl formal)	Ethylene dichloride, dioxan, glacial acetic acid, phenols	Aliphatic hydrocarbons
Poly(vinyl acetate)	Aromatic and chlorinated hydrocarbons, acetone, methanol, esters	Aliphatic hydrocarbons
Poly(vinyl esters):		
Poly(vinyl methyl ether)	Methanol, water, benzene	Alkalis, soluble salts, aliphatic hydrocarbons
Poly(vinyl ethyl ether)	Aromatic and chlorinated hydrocarbons, esters, alcohols, ketones	Water
Poly(vinyl butyl ether)	Aliphatic, aromatic and chlorinated hydrocarbons, ketones	Alcohols
Poly(vinyl alcohol)	Formamide, water	Ether, alcohols, esters, ketones, aliphatic and aromatic hydrocarbons
Poly(vinyl carbazole)	Aromatic and chlorinated hydrocarbons, tetrahydrofuran	Ether, alcohols, esters, hydrocarbons, ketones, carbon tetrachloride
Rubber	Chlorinated and aromatic hydrocarbons	Oxygen-containing solvents

N.B. For an extensive table see Kline.[6]

most cases it is only necessary to classify the sample and compare the polymeric component with known and standard samples.

With modern methods of information retrieval and pattern recognition involved in the interpretation of electric signals from modern analytical instruments it is generally superfluous to proceed further without resource to physical measurements. Only one main area needs to be considered further, and that is that associated with solubility. Solubility tests not only help to give a further indication of the probable classification, but also aid in the selection of suitable solvents for the preparation of solutions of the sample for further investigations.

There are therefore several methods of listing solvent/solute systems. At this point it is assumed that the probable classification has been deduced from previous tests and 'confirmation' and choice of a suitable solvent for solution preparation have equal priorities. The 'solubility table' (see Table 1.7) is therefore given in order to give preference to a 'classified sample'.

1.8. CONCLUSION

The present chapter is intended only to be a guide and aid to that which follows. The techniques of modern methods of analysis of polymers can be so refined and automated that chemical methods of identification and quantification which were necessary before the 1980s are now being superseded. Indeed, the whole area of analytical chemistry is so rapidly changing that it is probable that the 'General Analyst' will have to become the 'Analyst General' and direct the samples towards those 'Analytical Scientists' who have especial expertise in one selected area of instrumentation and related information retrieval.

REFERENCES

1. H. Mark, *Analyt. Chem.*, **20**, 104, 1948.
2. J. Haslam, H. A. Willis and D. C. M. Squirrell, *Identification and Analysis of Plastics*, 2nd edn., Iliffe, London, 1972.
3. J. Haslam, J. B. Hamilton and D. C. M. Squirrell, *Analyst*, **85**, 556, 1960.
4. R. Belcher, M. A. Leonard and T. S. West, *Talanta*, **2**, 92, 1959.
5. F. Feigl, *Spot Tests in Organic Chemistry*, 6th edn., Elsevier, New York, 1960, p. 103.
6. G. M. Kline (Ed.), *Analytical Chemistry of Polymers. III.*, Interscience Publishers, N.Y., 1962.

Chapter 2

MICROSTRUCTURAL ANALYSIS OF SYNTHETIC ORGANIC POLYMERS BY HIGH RESOLUTION NUCLEAR MAGNETIC RESONANCE

J. R. Ebdon

Department of Chemistry, University of Lancaster, UK

2.1. INTRODUCTION

The last twenty years have seen a spectacular growth in the use of nuclear magnetic resonance (NMR) for the microstructural analysis of polymers. The stimulus for this growth was the demonstration in the early 1960s that high resolution proton (^{1}H) NMR spectroscopy could be used to determine the tacticity of some acrylic and vinyl homopolymers and to assess the distributions of monomer units within some simple addition copolymers. Following these early experiments, progress was rapid, aided in large measure by the development of ^{1}H NMR spectrometers equipped with high-field superconducting magnets and capable of operating at radio-frequencies of 220, 300 or even 360 MHz. The advent of these instruments allowed the detection of more subtle features of polymer microstructure and extended the applicability of the ^{1}H NMR technique. By the end of the 1960s, the use of ^{1}H NMR for microstructural analysis of polymers was well established and the progress made through the decade was comprehensively reviewed by Bovey, the foremost contributor in the field.[1]

During the 1970s attention was turned away from ^{1}H NMR and towards ^{13}C NMR. Hitherto, ^{13}C NMR had been of only limited interest because of the long times needed to accumulate, with the aid of a computer, ^{13}C spectra of acceptable signal to noise ratio using continuous wave (CW) methods on materials containing the natural abundance (1·1 %) of the ^{13}C isotope. However, the development of Fourier transform (FT) techniques utilising short pulses of radiation led to about a hundred-fold reduction in the time required to acquire a ^{13}C NMR spectrum.

^{13}C NMR spectroscopy offers several advantages for the polymer chemist. The much larger range of chemical shifts observed for carbon nuclei in polymer chains (around 200 ppm) compared with those of protons

(13 ppm or less) means that often microstructural differences can be detected more readily by ^{13}C NMR than by ^{1}H NMR. Also, the backbones of most polymers are largely carbon based and so one would expect fundamental information regarding chain structure to be obtained most easily from the resonances of the carbon nuclei. However, the FT technique has attendant disadvantages, particularly in the care that must be exercised in choosing instrumental parameters if quantitative measurements are to be made. The rapid progress made in developing the FT ^{13}C NMR technique as applied to polymers can be gauged by comparing the spectra presented in a 1972 review by Mochel[2] with those presented by Randall[3] five years later.

The past few years have seen the continuing development of FT NMR with the method now extended to cover other nuclei of low abundance, e.g. ^{15}N. Also, it has recently become possible to obtain high resolution ^{13}C spectra on solid crystalline and amorphous polymers[4] (below the glass-transition temperature) and on cross-linked, intractable resins.[5] This last development promises to be of particular importance.

This chapter reviews principally the applications of high resolution NMR spectroscopy to the analysis of the microstructures of synthetic organic polymers. The chapter is intended to be illustrative rather than exhaustive with examples chosen to show the range of information which is available from NMR measurements. Comprehensive reviews of applications of NMR to polymer analysis are published annually by the Chemical Society[6] and biennially by the American Chemical Society.[7]

The applications of NMR to the determination of polymer conformation and to the measurement of relaxation parameters associated with macromolecular motion are not covered. Neither is broad-line NMR considered. It is assumed that the reader is familiar with the general principles of CW and FT NMR spectroscopy. There are many texts now available which deal with these principles[8-13] and several which consider features of particular relevance to polymer analysis.[1,3,14,15] Recently books dealing with the NMR of ^{15}N and other less abundant nuclei have appeared.[16,17]

2.2. IDENTIFICATION OF POLYMERS AND MEASUREMENTS OF COPOLYMER COMPOSITION

Many organic polymers can readily be identified from their ^{1}H and/or ^{13}C spectra. The major requirement for ^{1}H NMR is that they be soluble in a suitable solvent at the 5–10 % w/v level and for ^{13}C NMR that they at least swell in a solvent. The approach involved in polymer identification is essentially the same as that for small molecules and makes use either of tables of standard chemical shift data or of spectra of authenticated samples or suitable model compounds.

Line widths in polymer spectra are generally larger than those in the spectra of low molecular weight compounds owing to the dipolar interactions and restricted mobility that exist in viscous polymer media. In proton spectra of polymers, homonuclear coupling may not be easy to discern. To obtain the best spectra, solvents of low viscosity should be selected and there are advantages in recording spectra at above ambient temperature, to increase chain mobility, and at high field strengths, to improve resolution. However, it must be remembered that changes in temperature will bring about changes in average macromolecular confor- mations which can in turn produce marked changes in the appearance of a polymer NMR spectrum. For purely hydrocarbon polymers, ^{13}C spectra are likely to be more revealing than ^{1}H spectra. The resonance in the carbon spectra of hydrocarbon polymers can be assigned with the aid of published chemical shift additivity data.[18−21]

Copolymer compositions can often be determined from ^{1}H spectra recorded at relatively low magnetic fields. For example, Fig. 2.1 shows the 60 MHz ^{1}H spectrum of a copolymer of styrene and methyl methacrylate. The mole fraction (x) of styrene in the copolymer may be obtained with a probable error of within ± 2 % by measuring the area of the peak arising from the phenyl protons at around 7δ (A_{phenyl}) and comparing this with the total proton peak area (A_{total}). It can readily be shown that

$$x = \frac{8(A_{\text{phenyl}})}{5(A_{\text{total}})}$$

Similar procedures may be followed with other copolymers. The method requires that at least one resonance of a functional group in one of the monomer units should be adequately resolved from the rest of the spectrum. With CW ^{1}H spectra, care must be taken to ensure that the radiofrequency power is kept below saturation level if accurate quantitative results are to be obtained. With FT ^{13}C spectra, it is advisable to ensure a delay between pulses of at least five times the spin-lattice relaxation time, T_1, of the slowest relaxing carbon nucleus. Also, when recording ^{13}C spectra it is usual to remove ^{13}C–^{1}H couplings by broad-band irradiation over the proton resonances. For quantitative ^{13}C work it is best to employ appropriately gated decoupling so that nuclear Overhauser enhancements

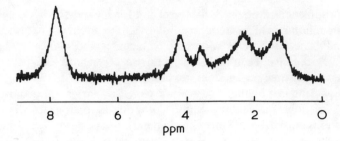

FIG. 2.1. ^{1}H spectrum of a random copolymer of styrene and methyl methacrylate recorded at 60 MHz on an approximately 10 % w/v solution in CDCl$_3$ at
35 °C. Chemical shifts are in ppm relative to TMS.

are suppressed.^{13}C NMR is likely to be a more reliable method of analysis
than is ^{1}H NMR for copolymers in which one of the monomer units
contains relatively small amounts of hydrogen, e.g. copolymers of styrene
and maleic anhydride. The compositions of multicomponent copolymers
also can often be determined by NMR.

2.3.   CONFIGURATIONAL SEQUENCES (TACTICITY) IN HOMOPOLYMERS

Vinyl and acrylic monomers of the types $CH_2{=}CHX$ and $CH_2{=}CXY$,
where X and Y are substituents, other than H, such as CH_3, Cl, CN or
$COOCH_3$, on polymerisation yield chains containing pseudoasymmetric
centres. Such chains may contain various configurational sequences.
Sequences of two monomer units (dyads) may have meso, m (I), or racemic,
r (II), configurations.

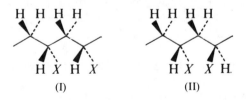

(I)   (II)

Sequences of three monomer units (triads) may have mm, mr (and rm) or
rr configurations, often given the names 'isotactic', 'heterotactic' and
'syndiotactic', respectively. The configurations of longer sequences (tetrads, pentads, hexads, etc.) can readily be described by extending the 'm'

and 'r' nomenclature system. Monomers of the type $CHX=CHY$ give chains containing threo-diisotactic, erythro-diisotactic and disyndiotactic placements.

The first polymer for which configurational information was obtained by NMR was poly(methyl methacrylate). Bovey and Tiers[22] demonstrated that the resonance of the α-methyl protons had three components which could be assigned in order of increasing field to monomer units at the centres of isotactic, heterotactic and syndiotactic triads. The β-methylene proton signals also could be used to give configurational information; the non-equivalent protons of the meso dyad give the typical four-line pattern of an AX spin system, whereas the equivalent protons of the racemic dyad give a broad central resonance. At higher fields, the α-methyl proton resonances show further fine structure characteristics of pentad effects (Fig. 2.2) and the methylene proton resonances show tetrad splittings.[23] The α-methyl proton resonances of poly(α-methyl styrene)[24] and of poly-methacrylonitrile[25] can similarly be used to assess the relative proportions of configurational triads; here too, higher fields allow pentad effects to be seen.[26]

For monosubstituted ethylenes, determining tacticity by ^1H NMR is a little less straightforward because of the complicating effects of the coupling between the β-methylene and α-methine protons. However, this

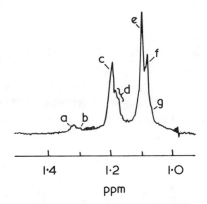

1·4   1·2   1·0

ppm

FIG. 2.2.   α-Methyl proton spectrum of a predominantly syndiotactic poly(methyl methacrylate) recorded at 220 MHz on an approximately 10%w/v solution in chlorobenzene at 135 °C. Chemical shifts are in ppm relative to TMS. Assignments to pentads containing meso (m) and racemic (r) placements are as follows: a = rmmr; b = rmmm; c = rmrr + mmrr; d = rmrm + mmrm; e = rrrr; f = mrrr and g = mrrm. (Reproduced from Bovey[1] with permission of Academic Press, Inc.)

complication can be circumvented by selective decoupling[27] or by selective deuteration of the monomer.[28] It may also be ameliorated by working at high field strengths at which the chemical shift differences between the resonances arising from monomer units in the various configurational sequences become significantly larger than the coupling constants.[29] This effect is illustrated for poly(vinyl chloride) in Fig. 2.3.

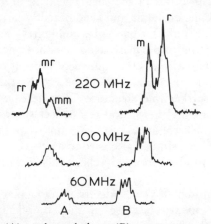

FIG. 2.3. Methine (A) and methylene (B) proton resonances of poly(vinyl chloride) in o-dichlorobenzene recorded at 60,100 and 220 MHz showing assignments to sequences containing meso (m) and racemic (r) placements. (Reproduced from Becconsall et al.[29] with permission of Marcel Dekker, Inc.)

^{1}H NMR is probably least suitable for determining the tacticities of purely hydrocarbon polymers such as polypropylene, although important information about the mechanism of propylene polymerisation with coordination catalysts has been obtained from studies of the ^{1}H spectra of polymers prepared from selectively deuterated monomer.[30] A wide variety of vinyl and acrylic homopolymers have now been examined by ^{1}H NMR. For many, configurational sequence information has been obtained fairly readily from the spectra, for others selective deuteration of the monomer has been necessary, and for some the spectra remain essentially unresolved.[1]

However, for the majority of vinyl and acrylic homopolymers, ^{13}C NMR is now considered to be the preferred technique for determining configurational microstructure. The reason for this lies in the much greater range of chemical shifts exhibited by carbon nuclei when compared to protons which usually allows discrimination between longer stereosequences. The availability of more detailed configurational information in turn permits the testing of more complex statistical models for the chain propagation processes. ^{13}C NMR has proved particularly useful for determining the tacticities of homopolymers for which the ^{1}H spectra are made complicated by proton–proton coupling. For example, the tacticity of polyacrylonitrile can be determined from the methine- and cyano-carbon signals, each of which is split into three peaks assignable to mm, mr (and rm) and rr triads.[31,32] Under conditions of higher resolution, the cyano-carbon signals are further split into nine peaks (Fig. 2.4) which may be assigned to configurational pentads.[33] Symmetry considerations

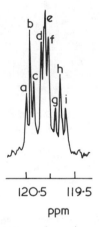

FIG. 2.4.   [13]C resonances of cyano-carbons in an essentially atactic polyacrylonitrile recorded at 25·2 MHz on a 10 % w/v solution in d₆-DMSO at 32 °C. Chemical shifts are in ppm relative to TMS. Assignments to pentads containing meso (m) and racemic (r) placements are as follows: a = mmmm; b = mmmr; c = rmmr; d = mrmm; e = mrmr + rrmm; f = rrmr; g = mrmm; h = rrrm and i = rrrr. (Reproduced from Balard *et al.*[33] with permission of Hüthig and Wepf Verlag.)

indicate that β-methylene carbon signals will usually show splittings characteristic of even 'ads' (dyads, tetrads, hexads, etc.) whilst α-carbons and the carbons of substituents attached to the α-position will show splittings characteristic of the odd 'ads' (triads, pentads, heptads, etc.).

Particular effort has been devoted to interpreting the ^{13}C spectra of polypropylenes.[3,14] Amorphous polypropylenes give rise to ^{13}C spectra showing fine structure for all three carbon resonances. The assignments of the components of these resonances can be carried out with the aid of spectra from pure isotactic and syndiotactic polymers,[34] from epimerised isotactic and syndiotactic polymers[35] and from suitable model compounds of known stereochemistries.[36] The procedures which have been followed by several groups and which have led to the assignments of the methyl carbon resonances shown in Fig. 2.5 have been reviewed by Randall[3] and Cunliffe.[14] The assignment of the β-methylene carbon resonances to configurational hexads has recently been completed by Schilling and Tonelli,[37] and by Zambelli et al.[38] The former group calculated the

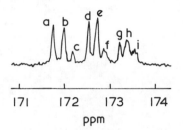

171   172   173   174
ppm

FIG. 2.5.   [13]C resonances of methyl carbons in an atactic polypropylene recorded at 25·2 MHz on a 3·5–7 % w/v solution in CS_2 at 135–140 °C. Chemical shifts are in ppm relative to CS_2. Assignments to pentads containing meso (m) and racemic (r) placements are as follows: a = mmmm; b = mmmr; c = rmmr; d = mmrm; e = mmrr + rmrm; f = rmrr; g = rrrr; h = mrrr and i = mrrm. (Reproduced from Zambelli et al.[34] with permission of American Chemical Society.)

chemical shifts expected for the carbon resonances of different configurational sequences using a rotational isomeric state model to determine the conformations of lowest energy and noting situations where γ-effects (three-bond gauche interactions) would lead to pronounced upfield chemical shifts. The latter group of workers assumed that the chemical shifts of the methylene carbons in polypropylene would be the same as those in 2,4,6,8,10,12,14,16-octamethylheptadecanes and derived the intensities for the various hexad stereosequences from the observed methyl pentad sequence intensities.

The methods applied to the analysis of the β-methylene carbon resonances in polypropylene promise to be of use in the assignment of the β-methylene carbon resonances in polystyrene[39] and poly(vinyl chloride).[40] Amongst the β-methylene carbon signals of polystyrene (Fig. 2.6), only that arising from the 'mm' tetrad can be assigned with certainty using the spectrum of a highly isotactic polymer. The remaining assignments are speculative and are based upon an assumed Bernoullian statistical behaviour.[41] For poly(vinyl chloride), an internally consistent

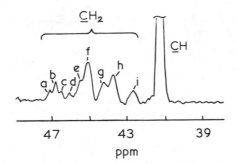

FIG. 2.6.   [13]C resonances of methylene carbons in an essentially atactic poly-styrene recorded at 25·2 MHz on a solution in 1,2,4-trichlorobenzene/d_6-benzene at 120 °C. Chemical shifts are in ppm relative to TMS. Assignments (some tentative) to tetrads and hexads containing meso (m) and racemic (r) placements are as follows: a = (mmrrr);   b = mmrrm;   c = (rmrrr);   d = (rmrrm);   e = (rmr);   f = mmr; g = (mrm); h = mmm and i = (rrr). (Reproduced from Randall[41] with permission of John Wiley and Sons, Inc.)

but empirical set of β-methylene carbon assignments has been made.[42] However, the pattern of the β-methylene carbon resonances for this polymer is very solvent dependent,[43] a fact which contributed to some early disagreement over assignments.[44,45] Some doubt about these assignments thus remains. The assignments to configurational pentads of the aromatic quaternary carbon signals in polystyrenes[46−48] and of the methine carbon signals in poly(vinyl chloride)[44] also need to be confirmed.

Other homopolymers for which tacticity information has been obtained by ^{13}C NMR include polyacrylates[49,50] and methacrylates,[51] poly(α-methyl styrenes),[46,52] poly(vinyl pyridines),[53−56] poly(N-vinyl carba-zole),[57] poly(vinyl acetates) and alcohols,[58−61] polypropylene oxides[62−64] and sulphides,[65−67] 1,2-polybutadiene,[68] and some poly(amino acids).[69]

The general conclusions about homopolymer tacticity that may be drawn from NMR studies are that for monomers possessing only one α-substituent, e.g. styrene, acrylonitrile, vinyl chloride and acrylates, the use of radical and cationic initiators leads to essentially atactic polymers in which the configurational sequence fractions are close to those predicted by Bernoullian statistics with a probability of meso placement, σ_m, close to 0·5. However, for monomers with two α-substituents, e.g. methyl methacrylate, methacrylonitrile and α-methylstyrene, syndiotactic-rich polymers are formed with these initiators and $\sigma_m < 0.5$. Highly isotactic or highly syndiotactic vinyl and acrylic polymers may be produced with suitable anionic initiators or coordination catalysts. In these cases, the distribution of configurational sequences is unlikely to follow simple Bernoullian statistics but may be approximated by Markov statistics of first or higher order.

2.4. MONOMER UNIT ISOMERISM

Several monomers when polymerised lead to polymers in which the repeat units exhibit geometrical isomerism. The most well known examples are the 1,3-dienes which can produce polymers containing cis-1,4-, trans-1,4-, 1,2- and (in the case of substituted dienes) 3,4- units. Relative amounts of 1,4- and 1,2- units can usually be determined from a simple ^{1}H spectrum at low field since only 1,2- units give rise to olefinic methylene resonances at around $4\cdot8\delta$ ppm. Cis-1,4- and trans-1,4- contents can be determined by ^{1}H NMR but usually only at high fields. For example, the cis- and trans-contents of polybutadiene have been determined from the olefinic methine resonances and aliphatic methylene resonances at 300 MHz.[70] For polyisoprenes, cis-1,4- and trans-1,4- contents can be determined from their respective methyl proton resonances at $1\cdot7\delta$ ppm and $1\cdot6\delta$ ppm.[71,72] Similar procedures are possible with polypentadiene[73] and poly-2,3-dimethylbutadiene.[74] However, for cis- and trans-1,4-polychloroprenes, the ^{1}H NMR spectra are rather similar,[75] although the two isomers can be differentiated at 220 MHz.[76]

Of interest also for diene polymers are the distributions of isomeric units along the polymer chains. Figure 2.7 shows that some of this additional information is available for polybutadiene from the ^{1}H spectra.[77] However, this sort of information is often better obtained by ^{13}C NMR. The distributions of 1,2- and 1,4- units in polybutadienes have been measured using ^{13}C NMR by Randall on hydrogenated polymers.[78] The hydrogenated polymers may be treated as branched alkanes for which all the necessary chemical shift information is available. The distributions of the 1,2- and 1,4- units in the polymers examined by Randall obeyed first-order Markov statistics. However, hydrogenation of polybutadienes removes the distinction between cis- and trans-1,4- units. Unhydrogenated polymers have been examined by several groups.[79-83] The most complete and self-consistent ^{13}C assignments are those of Suman and Werstler.[83] Of particular use are the C_1 and C_4 methylene carbon resonances which may be assigned to cis-1,4-, trans-1,4- and 1,2- units in various isomeric dyads and triads.

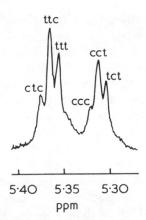

FIG. 2.7.   Decoupled olefinic methine proton resonances of a polyisoprene containing cis-1,4-(c) and trans-1,4-(t) units recorded at 300 MHz. Chemical shifts are in ppm relative to HMDS. Assignments to isomeric triads are indicated. (Reproduced from Santee et al.[77] with permission of John Wiley and Sons, Inc.)

The interpretation of the ^{13}C spectra of polyisoprenes and polychloroprenes is more difficult since 2-substituted dienes can be incorporated in adjacent 1,4- units with head-to-head (4,1—1,4) and tail-to-tail (1,4—4,1) linkages, as well as with the more usual head-to-tail (1,4—1,4) linkages. This particular feature is discussed later. Nevertheless, the ^{13}C spectra of a variety of polyisoprenes have been interpreted quantitatively and information about the distributions of isomeric units along the chains obtained.[84] The relative amounts of cis-1,4-, trans-1,4- and 3,4-units can be determined either from the C_1 (methylene) and C_3 (methine) signals or from the C_5 (methyl) signals. Approximate values for the fractions of dyads and triads containing these units can be obtained from the C_1 and C_3 signals also. The spectra of hydrogenated polyisoprenes also allow the distributions of 1,4- and 3,4- units to be determined but the distinction between cis-1,4- and trans-1,4- is of course removed.[85,86] It has been found that, for a series of polyisoprenes prepared with a BuLi/Et$_2$O catalyst, the distributions of isomeric units along the chain are essentially random indicating that the structure of the terminal monomer unit at the 'living' end of the chain is little affected by the structure of the penultimate monomer unit.[84] In these particular polyisoprenes, no head-to-head and tail-to-tail linkages are evident.

The ^{13}C spectra of some mainly trans-1,4-polychloroprenes have been interpreted by Coleman et al.[87] Lines in the olefinic and methylene regions of the spectra have been assigned to dyad and triad sequences involving the predominant trans-1,4- units. Other lines are associated with cis-1,4-, 1,2-, isomerised 1,2- and 3,4- units. The spectra also show clear evidence of significant amounts of head-to-head and tail-to-tail placements.

Other diene homopolymers examined by ^{13}C NMR include poly-1,3-pentadiene,[88] poly-1-phenylbutadiene[89] and poly-2,3-dimethylbutadiene.[90] The spectra of poly-1,3-pentadienes can be used to determine relative amounts of trans-1,4-, cis-1,4- and 1,2-enchainment and also to determine the tacticities of the head-to-tail linked 1,4-sequences. The spectra of poly-1-phenylbutadienes and poly-2,3-dimethylbutadienes give some isomer sequence information as well as relative amounts of trans-1,4-, cis-1,4- and, in the case of poly-1-phenylbutadiene, 3,4- enchainment.

^{13}C NMR has proved valuable also in the determination of isomer distributions in polypentenamers,[91,92] in other polymers produced by the ring-opening of cyclic alkenes[93–95] and in cationic polymerisation of higher olefins.[96] ^{15}N NMR has been used recently to investigate the structures of some polyamides and polylactams.[97]

2.5. LINKAGE ISOMERISM

The polymerisation of substituted olefins and dienes normally gives rise to chains in which the monomer units are linked in an almost exclusively head-to-tail fashion. This arises from the dipolar interactions between the monomer unit at the chain end and the incoming monomer molecule and also because of the lower activation energy associated with the propagation step that places the active centre (radical, cation or anion) at the most stable position on the monomer unit, usually, in the case of olefins, at the most substituted position. Coordination catalysts also tend to coordinate the monomer in one particular orientation and therefore to produce predominantly head-to-tail linked chains. However, for some monomers and/or catalysts, the orienting influences are relatively small and significant amounts of the alternative head-to-head and tail-to-tail linkages can arise. One of the first polymers in which these possibilities were recognised, through the use of NMR spectroscopy, was poly(vinylidene fluoride).[98] Figure 2.8 shows the ^{19}F spectrum of a poly(vinylidene fluoride). In addition to the predominant peak arising from the CF_2 units in regular head-to-tail sequences there can be seen smaller peaks at higher fields which are characteristic of head-to-head and tail-to-tail sequences. A similar

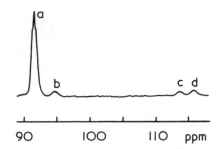

FIG. 2.8. ^{19}F spectrum of a poly(vinylidene fluoride) recorded at 56·4 MHz on a 25 % w/v solution in N,N-dimethylacetamide. Chemical shifts are in ppm relative to $CFCl_3$. Assignments are as follows: a = —CF_2—CH_2—$C\underline{F}_2$—CH_2—CF_2—; b = —CH_2—CH_2—$C\underline{F}_2$—CH_2—CF_2—; c = —CF_2—CH_2—$C\underline{F}_2$—CF_2—CH_2—; d = —CH_2—CH_2—$C\underline{F}_2$—CF_2—CH_2—. (Reproduced from Wilson[98] with permission of John Wiley and Sons, Inc.)

phenomenon can be recognised in the high field proton spectra of polychloroprenes in which the large peak arising from the methylene groups in 1,4—1,4 sequences is flanked by two smaller peaks characteristic of 4,1—1,4 and 1,4—4,1 sequences.[75] These structural irregularities are seen even more clearly in the ^{13}C spectra of polychloroprenes in which peaks associated with both *cis*-1,4- and *trans*-1,4- units in head-to-head and tail-to-tail linkages can be separately identified.[87,99] In commerical polychloroprenes, the amounts of these irregularities increase with increasing polymerisation temperature as expected.[100] The same pattern of resonances for head-to-head and tail-to-tail linkages is observed also in ^{13}C spectra of predominantly 1,4-polyisoprenes prepared with radical initiators.[84,101,102] For these polymers, the proportions of *trans*-1,4- units and *cis*-1,4- units in head-to-head linkages can be obtained from the areas of the C_1 methylene resonances at 38·6 and 31·4 ppm, respectively, whilst tail-to-tail linkages involving both *cis*- and *trans*- units can be estimated from the C_4 methylene resonances between 28·4 and 28·8 ppm. The resonance corresponding to the C_4 methylenes in tail-to-tail linkages involving *trans*-1,4- units can be unequivocally assigned since this linkage occurs also in the natural product squalene.

Peaks arising from the inversion of monomer units have been detected also in the ^{13}C spectra of polypropylenes, particularly those prepared with homogeneous vanadium-based aspecific and syndiospecific catalysts.[103,104] To the high field side of the main methyl carbon resonances can be found a group of small peaks, several of which can be assigned to the methyl carbons of propylene units in head-to-head linkages. There is some disagreement over the probable positions of methyl resonances arising from tail-to-tail linkages but on the basis of one possible set of assignments, Doi[105] has concluded that the sequence distributions of inverted propylene units can be accounted for in terms of first-order Markov statistics, i.e. head-to-head and tail-to-tail linkages tend to occur in isolated positions along the chain rather than in a random distribution.

2.6. MONOMER SEQUENCES AND CONFIGURATIONAL SEQUENCES IN COPOLYMERS

The main interest in examining the microstructure of a copolymer is usually to determine the statistical nature of the process governing the incorporation of the various monomer units into the copolymer chains. This can be achieved by measuring the monomer sequence distribution or some features of it. Often it is found that the frequency of a nuclear magnetic resonance in one of the monomer units is sensitive to the natures of the neighbouring monomer units. Thus in a binary copolymer consisting of monomer units A and B it is often possible to distinguish, for example, between the triads AAA, AAB and BAB. In favourable cases, longer sequences may be distinguishable. As with configurational sequences, β-methylene proton and carbon resonances usually give information about the even 'ads' (dyads, tetrads, hexads, etc.), whereas α-methine proton and carbon, α-quaternary carbon and side-group proton and carbon resonances are sensitive to odd 'ads' (triads, pentads, heptads, etc.).

The importance of measuring monomer sequence distributions lies in the information which this can sometimes provide about the mechanism of copolymerisation. The simplest realistic model that can be applied to a binary copolymerisation is the so-called terminal model, in which the compositions and monomer sequence distributions of the copolymers can be calculated from the compositions of the monomer feeds using only two variable parameters or reactivity ratios (first-order Markov statistics). Deviations from first-order Markov statistics may indicate (a) the dependence of chain reactivity upon the nature of the penultimate, or some more remote, monomer unit, (b) the involvement of comonomer complexes in propagation, or (c) the significance of depropagation reactions.

The easiest type of copolymer to analyse by NMR from the point of view of obtaining information about monomer sequences is one in which neither monomer unit can exhibit geometrical or stereoisomerism. An example of such a system is isobutylene(I)–vinylidene chloride(V). Fischer *et al.*[106] showed that the methylene proton resonances in these copolymers occurred in three main regions which could be assigned, in order of increasing field, to methylene groups in VV, IV + VI and II dyads. The proportion or dyad fractions calculated from the spectra were in reasonable agreement with those calculated from reactivity ratios. The methylene resonances showed further fine structure attributable to tetrad effects. The similar ethylene–vinylidene chloride system has also been examined by NMR.[107,108]

An alternative approach to simplifying monomer sequence analysis in copolymers is to prepare copolymers in which all of the monomer units have asymmetric or pseudoasymmetric centres of the same type. For example, Klesper has prepared cosyndiotactic copolymers of methyl methacrylate and methacrylic acid by partly hydrolysing syndiotactic poly(methyl methacrylates).[109] Compositional triad fractions may be obtained from the methyl proton resonances. With the aid of these spectra, those of copolymers with different configurational sequences can be interpreted.[110] A similar approach has been followed with the ^{13}C spectra of these copolymers.[111]

Other situations where monomer sequence information may be fairly readily obtained from NMR spectra are those where only one of the monomers exhibits stereoisomerism, especially if this monomer is present in the copolymers in only small amounts. An example of great commercial significance is ethylene–propylene. Most rewarding for study are the ^{13}C spectra of these copolymers for which most complete analyses have been presented by Carman *et al.*[112] and by Randall.[113] The information available from these spectra has recently been reviewed.[114] Other copolymers of this general type examined by ^{1}H and/or ^{13}C NMR include ethylene–vinyl chloride,[108] ethylene–vinyl acetate,[115–119] ethylene–vinyl alcohol,[120,121] ethylene–vinyl formate,[122] ethylene–methyl acrylate,[123] ethylene–glycidyl acrylate,[124] tetrafluoroethylene–propylene,[125] vinylidene chloride–methacrylonitrile,[126–128] vinylidene chloride–acrylonitrile,[129] vinylidene chloride–vinyl acetate,[130] isobutylene–chlorotrifluoroethylene[131] and vinylidene fluoride–hexafluoropropylene.[132] The fluorocarbon systems have been examined also by ^{19}F NMR.[131–133]

For vinylidene chloride (V)–methacrylonitrile (M) copolymers, information pertinent to monomer sequence may be obtained from both high field ^{1}H spectra and from ^{13}C spectra.[127,128] Figure 2.9 depicts the methylene proton region at 220 MHz for such a copolymer in solution in

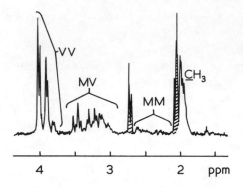

Fig. 2.9.   Methylene proton resonances of a vinylidene chloride (V)–methacrylo-nitrile(M) copolymer containing 71 mol % V recorded at 220 MHz on an approxi-mately 10 % w/v solution in d_6-acetone at 40 °C, showing assignments to compositional (VV, MV and MM) dyads. Chemical shifts are in ppm relative to TMS. (Reproduced from Suggate[127] with permission of Hüthig and Wepf Verlag, Basel.)

perdeuteroacetone. The methylene peaks occur in three main regions; these can be assigned, in order of increasing field, to methylene groups at the centres of VV, MV + VM and MM dyads. The MV + VM and MM peaks are further split in a complex way reflecting both the effects of longer monomer sequences and also of different methacrylonitrile unit con-figurations. However, the further splitting of the VV peaks arises only from monomer sequence effects, and from the areas of these, the amounts of the VV centred tetrad (and some hexad) fractions can be obtained. The ^{13}C spectra are even more revealing with nearly all the carbon resonances showing splittings arising from monomer sequence and configurational sequence effects (Fig. 2.10). Methacrylonitrile centred triad fractions may be obtained from the —$\underline{C}$N and —$\underline{C}$(CH$_3$)CN resonances, and vinylidene chloride centred triad and pentad fractions may be obtained from the —$\underline{C}$Cl$_2$ resonances. Measurements of these sequence fractions for radical copolymers prepared in several solvents indicate that the copolymerisation reactivity ratios are influenced by the nature of the solvent and the composition of the comonomer feed.

For copolymers in which both monomer units possess an asymmetric or pseudoasymmetric centre, the question of cotacticity arises. For such copolymers, there is an interest in determining whether or not the generation of the cotactic sequences is a simple statistical process, i.e. can the process be described in terms of just a single parameter, $\sigma_{m_{AB}}$, the

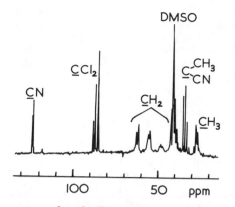

FIG. 2.10. ^{13}C spectrum of a vinylidene chloride–methacrylonitrile copolymer containing 64 mol % vinylidene chloride recorded at 20 MHz on an approximately 20 % w/v solution in DMSO/d$_6$-DMSO at 47 °C. Chemical shifts are in ppm relative to TMS. (Reproduced from Suggate[128] with permission of Hüthig and Wepf Verlag, Basel.)

probability of forming the compositional dyad AB (or BA) with a meso configuration? Many such copolymerisations have been examined both by ^{1}H and ^{13}C NMR. A simple approach is to prepare, initially at least, copolymers with a known and regular monomer sequence distribution. For such a system, the splitting of NMR peaks arising from configurational differences can then be more readily discerned. This approach has been followed for α-methylstyrene (A)–methacrylonitrile (M) copolymers by Elgert and Stützel.[134] They prepared regular alternating copolymers of these monomers, using ethyl aluminium sesquichloride as the regulating agent. Of particular value for the configurational analyses of these copolymers are the α-methylstyrene and methacrylonitrile methyl, and α-methylstyrene C$_1$ ring carbon resonances. The components of these peaks have been assigned to A and M centred configurational triad and A centred configurational pentad sequences, respectively. The configurational fractions obtained from the peaks fit a Bernoullian process with $\sigma_{m_{AM}} = 0.34$. This value is intermediate between those deduced for typical poly(α-methylstyrenes) ($\sigma_{m_{AA}} = 0.20$–0.31)[135] and typical poly(methacrylonitriles) ($\sigma_{m_{MM}} = 0.45$).[136] Thus the comonomer units do not appear to interact in any special way during the copropagation steps. Other systems for which cotacticity parameters have been obtained from ^{13}C spectra include methyl methacrylate (M')–α-methacrylophenone (P) for which $\sigma_{m_{M'P}} = 0.40$,[137] styrene (S)–methyl methacrylate (M') for which $\sigma_{m_{M'M'}} = 0.23$, $\sigma_{m_{SS}} = 0.29$

and $\sigma_{m_{M'S}} = 0\cdot50$,[138] and acrylonitrile (A')–methyl methacrylate (M') for which $\sigma_{m_{A'A'}} = 0\cdot5$ and $\sigma_{m_{M'A'}} = 0\cdot5$.[139] For the last two systems, the generation of cotactic sequences thus appears to be an entirely random process. Some other vinyl and acrylic copolymers that have been studied by ^{1}H and/or ^{13}C NMR include: styrene–methacrylonitrile,[140] methyl acrylate–methacrylonitrile,[141] styrene–methacrylic acid,[142] α-methyl-styrene–methyl methacrylate,[143] methyl methacrylate–methyl acrylate,[144] styrene–acrylonitrile,[145–147] vinyl alcohol–vinyl acetate,[148] methyl methacrylate–vinyl chloride[149] and vinyl chloride–propylene.[150,151]

Copolymers based on one or more diene monomers give particularly complicated NMR spectra in which the effects of monomer sequences, configurational sequences, isomeric sequences and linkage isomerism may all be seen. The ^{13}C spectra of some essentially random methyl methacrylate–butadiene copolymers have recently been partly interpreted.[152] It is reported that the isomer distributions of the butadiene units in the copolymers are influenced by the presence of the methacrylate units; copolymers rich in methyl methacrylate have a slightly greater proportion of *trans*-1,4-butadiene units than those which are rich in butadiene. Other diene-based copolymers for which ^{1}H and/or ^{13}C NMR analyses have been published include styrene–butadiene,[153–158] butadiene–acrylonitrile,[159–162] butadiene–methacrylonitrile,[163,164] butadiene–propylene,[165,166] isoprene–methyl methacrylate[167] and chloroprene–methyl methacrylate.[168–172] Sequence analyses of some polysulphones[173–175] and polyperoxides[176] by NMR indicate the importance of depropagation reactions in these systems.

The use of NMR for copolymer characterisation is not confined to addition copolymers; Kricheldorf and coworkers have applied ^{13}C NMR to the analysis of a variety of copolyamides, copolylactams and copolyesters.[177,178] The same group of workers has also demonstrated that ^{15}N NMR is suitable for the sequence analysis of certain polypeptides.[179] It is found for these polymers that the ^{15}N resonances of the amide groups are sensitive to the natures of the intervening groups.

2.7.   MINOR STRUCTURES: CHAIN BRANCHING AND END-GROUPS

Chain branching plays an important part in determining the properties of low density polyethylene. The extent and nature of the branching in this polymer is most readily studied by ^{13}C NMR.[180–185] The ^{13}C spectrum of a low density polyethylene shows several small peaks around the main methylene carbon resonance at 30 ppm (Fig. 2.11). Using known additivity parameters for linear and branched hydrocarbons, it is possible to assign most of these resonances. There is general agreement that the spectra indicate that the major type of branches in low density polyethylene are butyl groups as would be expected on the basis of the Roedel 'back-biting' mechanism.[186] However, there has been some disagreement over the

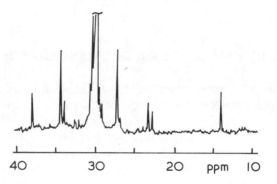

FIG. 2.11.   [13]C spectrum of a low density polyethylene recorded at 25 MHz on a 20 % w/v solution in 1,2,4-trichlorobenzene at 100 °C, showing minor resonances associated with chain branches. Chemical shifts are in ppm relative to TMS. (Reproduced from Bovey et al.[182] with permission of American Chemical Society.)

amounts of ethyl branches in such polymers and also over the nature of the longer branches. Bovey and coworkers have shown that ^{13}C NMR may be used also to examine chain branching in poly(vinyl chloride) but that it is first necessary to simplify the spectra by reducing the polymers with lithium aluminium hydride.[187] On reduction of the polymer, the carbons of the main chains give a single sharp peak whilst the branch carbons give distinct smaller resonances. It has been concluded that the main branches in poly(vinyl chloride) are methyl groups at a concentration of about 3 per 1000 carbon atoms.[187,188]

Resonances arising from the end-groups of polymer chains can often be seen provided that the molecular weights of the polymers are sufficiently low or that the materials have been artificially enriched with [13]C. Zambelli and coworkers[104,109] have assigned, to the methyl groups of saturated end-groups, small peaks to the high and low field side of the main methyl carbon resonances in polypropylene. Caraculacu and colleagues[190] have identified unsaturated end-groups in poly(vinyl chloride) by Fourier transform ^{1}H NMR. A more recent study[191] by ^{13}C NMR of low molecular weight species extracted from poly(vinyl chloride) with methanol has revealed the presence of unsaturated end-groups of the types —CHClCH$_2$CH= CHCH$_2$OCH$_3$ and —CH$_2$CHClCH=CHCH$_2$Cl, and saturated end-groups with the structures —CHClCH$_2$Cl and —CH$_2$CH$_2$Cl. End-, middle- and branching-groups in polysiloxanes can easily be detected and relative concentrations have been measured by Horn et al.[192] using ^{29}Si NMR.

2.8.   THERMOSETTING RESINS AND SOLID POLYMERS

The structures of thermosetting resins below the gel point are amenable to study by the usual NMR methods. ^{13}C NMR in particular has proved to be useful in the analysis of phenol-formaldehyde,[193–195] urea-formaldehyde[194,196–200] and melamine-formaldehyde resins.[194,201,202] The ^{13}C NMR spectrum of a low molecular weight urea-formaldehyde resin is shown in Fig. 2.12. Various methylene carbon resonances are visible which may be assigned, with the aid of model compounds, to methylol

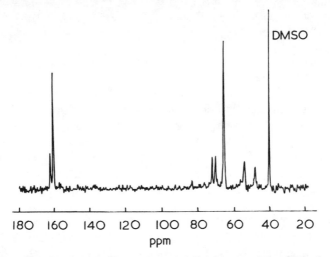

FIG. 2.12. ^{13}C spectrum of a commercial liquid urea-formaldehyde resin recorded at 20 MHz on an approximately 20 % w/v solution in D$_2$O/DMSO at 47 °C. Chemical shifts are in ppm relative to TMS. (Reproduced from Ebdon and Heaton[196] with permission of IPC Business Press Ltd.)

groups (65·5 ppm and 72·1 ppm), methylene ether linkages in linear chains and at branch points (70·0 ppm and 76·5 ppm) and methylene linkages in linear chains and at branch points (47·7 ppm and 54·3 ppm). The carbonyl resonances between 160 and 165 ppm are almost equally informative. With melamine-formaldehyde resins, the most useful resonances are those associated with the azine ring carbons which show splittings that depend upon the pattern of substitution at the melamine amine groups.[201]

Above their glass transition temperature (T_g), many polymers give ^{13}C spectra of acceptable resolution. The reason for this lies in the increased macromolecular mobility that exists above T_g which reduces the severity of the dipolar broadening otherwise observed for solid polymers. Solid samples of natural rubber at ambient temperatures give spectra in which the signals from all five of the carbon atoms in the repeat unit are clearly visible.

However, the analysis of thermosetting resins at temperatures beyond the gel point and of other solid polymers below their glass transition temperatures cannot be accomplished by normal NMR techniques. Even for a 'dilute nucleus' such as ^{13}C, dipolar interactions (in this case between ^{13}C and ^{1}H nuclei) are considerable and lead to broad ill-defined resonance lines. However, much of this broadening can be removed by strong, dipolar decoupling using powers about a hundred times stronger than that used for the scalar decoupling of ^{13}C from ^{1}H in the normal solution experiments. By this method, Schaefer et al.[4] were able to produce a ^{13}C spectrum of reasonable resolution from a solid sample of poly(methyl methacrylate). However, the long relaxation times of ^{13}C nuclei (of the order of a minute) in a solid matrix make for prohibitively long spectral accumulation times. This problem can be overcome by employing a multiple pulse sequence, which establishes a cross polarisation between the carbon spins and the more numerous proton spins. In this way, the time taken to repolarise the carbon spins is governed by the much shorter times for proton spin–lattice relaxation. Even so, the resultant spectra are likely still to exhibit rather broad lines, because of residual dipolar broadening and chemical shift anisotropy. Most of this residual broadening can be removed by spinning the sample at the 'magic angle' of 54·7° to the static field axis. ^{13}C spectra of many solid polymers have now been recorded with the aid of these devices, including those of polystyrene,[4] polyethylene,[203,204] cured and filled cis-1,4-polyisoprene,[205] aromatic polyesters,[206] epoxy resins,[207] polysulphone,[208] polyether sulphone,[208] polyphenylene oxide,[208] and cellulose.[209,210] From experiments on solid polymers involving cross-polarisation can be obtained a new relaxation parameter, $T_{1\rho}$, the rotating frame relaxation time. This relaxation time relates to relaxation processes occurring in the polymer with frequencies of the order of kilohertz rather than megahertz, the frequencies associated with the normal spin–lattice relaxation time, T_1. $T_{1\rho}$ values have been measured for several solid polymers and a remarkable correlation has been shown to exist between these values and the impact strengths of the polymers concerned.[208] It is believed that the relaxation processes measured by $T_{1\rho}$ are those involved in the relaxation of mechanical stresses in polymers.

2.9. CHEMICAL REACTIONS ON POLYMERS

NMR methods are ideally suited to following the progress of many chemical reactions involving polymers. Many reactions which occur in solution especially can be monitored by NMR and could with advantage be carried out *in situ* in the spectrometer. The advantages of NMR methods are that they are quick, non-destructive and accurate. Difficulties could however be experienced in trying to follow fast reactions by FT NMR; the time taken to sample the spectrum must be short compared to the duration of the reaction. The literature on studies of polymer reactions by NMR is legion. Some recent examples concern the use of ^{13}C NMR to characterise cellulose nitrates[211] and acetates.[212] It is found for the former system that the carbon signals give information not only about the preferred sites of nitration but also about the sequence distributions of reacted cellulose units. ^{13}C NMR has been used recently also to follow the chlorination of polyethylene[213] and poly(vinyl chloride).[214] For polyethylene in solution, the spectra indicate a tendency for alternating chlorinated structures to be formed. ^{31}P NMR has proved of value, together with ^{13}C NMR, in a study of the thermolysis of polydimethoxyphosphazene.[215]

REFERENCES

1. F. A. Bovey, *High Resolution NMR of Macromolecules*, Academic Press, New York, 1972.
2. V. D. Mochel, *J. Macromol. Sci.-Revs. Macromol. Chem.*, **C8**, 289, 1972.
3. J. C. Randall, *Polymer Sequence Determination, Carbon-13 NMR Method*, Academic Press, New York, 1977.
4. J. Schaefer, E. O. Stejskal and R. Buchdahl, *Macromolecules*, **8**, 291, 1975.
5. H. A. Resing and W. B. Moniz, *Macromolecules*, **8**, 560, 1978.

6. F. Heatley, *Specialist Periodical Reports. Nuclear Magnetic Resonance*, Vol. 8, R. J. Abraham (Ed.), Chemical Society, London, 1979, Ch. 10.

7. See for example, *Anal. Chem.*, **51**, 293R, 1979.

8. J. W. Emsley, J. Feeney and L. H. Sutcliffe, *High Resolution Nuclear Magnetic Resonance Spectroscopy*, Pergamon Press, New York, 1964.

9. J. A. Pople, W. G. Schneider and H. J. Bernstein, *High Resolution Nuclear Magnetic Resonance*, McGraw-Hill, New York, 1959.

10. J. B. Stothers, *Carbon-13 NMR Spectroscopy*, Academic Press, New York, 1972.

11. E. Breitmaier and W. Voelter, ^{13}C *NMR Spectroscopy, Monographs in Modern Chemistry*, Vol. 5, Verlag Chemie, 1974.

12. G. C. Levy and G. L. Nelson, *Carbon-13 Nuclear Magnetic Resonance for Organic Chemists*, Wiley-Interscience, New York, 1972.

13. T. C. Farrar and E. D. Becker, *Pulse and Fourier Transform NMR*, Academic Press, New York, 1971.

14. A. V. Cunliffe, in *Developments in Polymer Characterisation—1*, J. V. Dawkins (Ed.), Applied Science Publishers Ltd, London, 1978, Ch. 1.

15. J. Schaefer and E. O. Stejskal, *Topics in Carbon-13 NMR Spectroscopy*, Vol. 3, G. C. Levy (Ed.), Wiley-Interscience, New York, 1979, Ch. 4.

16. G. C. Levy and R. L. Lichter, *Nitrogen-15 Nuclear Magnetic Resonance Spectroscopy*, Wiley-Interscience, New York, 1979.

17. R. K. Harris and B. E. Mann (Eds.), *NMR and the Periodic Table*, Academic Press, London, 1978.

18. D. M. Grant and E. G. Paul, *J. Amer. Chem. Soc.*, **86**, 2984, 1964.

19. L. P. Lindeman and J. Q. Adams, *Anal. Chem.*, **43**, 1245, 1971.

20. C. J. Carman, A. R. Tarpley and J. G. Goldstein, *Macromolecules*, **6**, 719, 1973.

21. J. C. Randall, *J. Polym. Sci., Polym. Phys. Ed.*, **13**, 901, 1975.

22. F. A. Bovey and G. V. D. Tiers, *J. Polym. Sci.*, **44**, 173, 1960.

23. H. L. Frisch, C. L. Mallows, F. Heatley and F. A. Bovey, *Macromolecules*, **1**, 533, 1968.

24. S. Brownstein, S. Bywater and D. J. Worsfold, *Makromol. Chem.*, **48**, 127, 1961.

25. A. Yamada and M. Yanagita, *J. Polym. Sci., B*, **9**, 103, 1971.

26. T. Suzuki, S. Koshiro and Y. Takegami, *Polymer*, **14**, 549, 1973.

27. F. A. Bovey, E. W. Anderson, D. C. Douglass and J. A. Manson, *J. Chem. Phys.*, **39**, 1199, 1963.

28. T. Yoshino and J. Komiyama, *J. Polym. Sci. B*, **3**, 311, 1965.

29. J. K. Becconsall, P. A. Curnuck and M. C. McIvor, *Applied Spectroscopy Reviews*, Vol. 4, E. G. Braeme (Ed.), Marcel Dekker, Inc., New York, 1971, p. 334.

30. F. C. Stehling, *J. Polym. Sci.*, **A2**, 1815, 1964.

31. J. Schaefer, *Macromolecules*, **4**, 105, 1971.

32. Y. Inoue and A. Nishioka, *Polym. J.*, **3**, 149, 1972.

33. H. Balard, H. Fritz and J. Meybeck, *Makromol. Chem.*, **178**, 2393, 1977.

34. A. Zambelli, D. E. Dorman, A. I. R. Brewster and F. A. Bovey, *Macromolecules*, **6**, 925, 1973.

35. F. C. Stehling and J. R. Knox, *Macromolecules*, **8**, 595, 1975.

36. A. Zambelli, P. Locatelli, G. Bajo and F. A. Bovey, *Macromolecules*, **8**, 687, 1975.
37. F. C. Schilling and A. E. Tonelli, *Macromolecules*, **13**, 270, 1980.
38. A. Zambelli, P. Locatelli, A. Provasoli and D. R. Ferro, *Macromolecules*, **13**, 267, 1980.
39. A. E. Tonelli, *Macromolecules*, **12**, 252, 1979.
40. A. E. Tonelli, F. C. Shilling, W. H. Starnes, Jr., L. Shepherd and I. M. Plitz, *Macromolecules*, **12**, 78, 1979.
41. J. C. Randall, *J. Polym. Sci.*, *Polym. Phys. Ed.*, **13**, 889, 1975.
42. C. J. Carman, *Macromolecules*, **6**, 725, 1973.
43. I. Ando, Y. Kato and A. Nishioka, *Makromol. Chem.*, **177**, 2759, 1976.
44. C. J. Carman, A. R. Tarpley, Jr. and J. H. Goldstein, *Macromolecules*, **4**, 445, 1971.
45. Y. Inoue, I. Ando and A. Nishioka, *Polym. J.*, **3**, 246, 1972.
46. Y. Inoue, A. Nishioka and R. Chûjô, *Makromol. Chem.*, **156**, 207, 1972.
47. K. Matsuzaki, T. Uryu, K. Osada and T. Kawamura, *Macromolecules*, **5**, 816, 1972.
48. B. Jasse, F. Lauprêtre and L. Monnerie, *Makromol. Chem.*, **178**, 1987, 1977.
49. K. Matsuzaki, T. Kanai, T. Kawamura, S. Matsumoto and T. Uryu, *J. Polym. Sci.*, *Polym. Chem. Ed.*, **11**, 961, 1973.
50. K. Hatada, T. Kitayama and R. W. Lenz, *Makromol. Chem.*, **179**, 1951, 1978.
51. Y. Inoue, A. Nishioka and R. Chûjô, *Polym. J.*, **2**, 535, 1971.
52. K.-F. Elgert, R. Wicke, B. Stützel and W. Ritter, *Polymer*, **16**, 465, 1975.
53. G. M. Lukovkin, O. P. Komarova, V. P. Torchilin and Yu. E. Kirsch, *Vysokomol. Soedin.*, **15A**, 443, 1973.
54. M. Brigodiot, H. Cheradame, M. Fontanille and J. P. Vairon, *Polymer*, **17**, 254, 1976.
55. K. Matsuzaki, T. Kanai, T. Matsubara and S. Matsumoto, *J. Polym. Sci.*, *Polym. Chem. Ed.*, **14**, 1475, 1976.
56. K. Matsuzaki, T. Matsubara and T. Kanai, *J. Polym. Sci.*, *Polym. Chem. Ed.*, **15**, 1573, 1977.
57. T. Kawamura and K. Matsuzaki, *Makromol. Chem.*, **179**, 1003, 1978.
58. Y. Inoue, R. Chûjô and A. Nishioka, *J. Polym. Sci.*, *Polym. Phys. Ed.*, **11**, 393, 1973.
59. T. K. Wu and D. W. Ovenall, *Macromolecules*, **6**, 582, 1973.
60. T. K. Wu and D. W. Ovenall, *Macromolecules*, **7**, 776, 1974.
61. T. K. Wu and M. L. Sheer, *Macromolecules*, **10**, 529, 1977.
62. N. Oguni, K. Lee and H. Tani, *Macromolecules*, **5**, 819, 1972.
63. T. Uryu, H. Shimazu and K. Matsuzaki, *J. Polym. Sci.*, *Polym. Letts. Ed.*, **11**, 275, 1973.
64. W. Lapeyre, H. Cheradame, N. Spassky and P. Sigwalt, *J. Chim. Phys.*, **70**, 838, 1973.
65. S. Boileau, H. Cheradame, P. Guerin and P. Sigwalt, *J. Chim. Phys.*, **69**, 1420, 1972.
66. S. Boileau, H. Cheradame, W. Lapeyre, L. Sousselier and P. Sigwalt, *J. Chim. Phys.*, **70**, 879, 1973.
67. P. Guerin, S. Boileau, F. Subira and P. Sigwalt, *Europ. Polym. J.*, **11**, 337, 1975.

68. K.-F. Elgert, G. Quack and B. Stutzel, *Makromol. Chem.*, **175**, 1955, 1974.
69. W. E. Hull and H. R. Kricheldorf, *J. Polym. Sci.*, *Polym. Lett. Ed.*, **16**, 215, 1978.
70. E. R. Santee, V. D. Mochel and M. Morton, *J. Polym. Sci.*, *Polym. Lett. Ed.*, **11**, 449, 1973.
71. H. Y. Chen, *Anal Chem.*, **34**, 1134 and 1793, 1962.
72. M. A. Golub, S. A. Fuqua and N. S. Bhacca, *J. Amer. Chem. Soc.*, **84**, 4981, 1962.
73. D. H. Beebe, C. E. Gordon, R. N. Thudium, M. C. Throckmorton and T. L. Hanlon, *J. Polym. Sci.*, *Polym. Chem. Ed.*, **16**, 2285, 1978.
74. D. Blondin, J. Regis and J. Prud'homme, *Macromolecules*, **7**, 187, 1974.
75. R. C. Ferguson, *J. Polym. Sci.*, **A2**, **11**, 4735, 1964.
76. T. Okada and T. Ikushige, *J. Polym. Sci.*, *Polym. Chem. Ed.*, **14**, 2059, 1976.
77. E. R. Santee, V. D. Mochel and M. Morton, *J. Polym. Sci.*, *Polym. Lett. Ed.*, **11**, 453, 1973.
78. J. C. Randall, *J. Polym. Sci.*, *Polym. Phys. Ed.*, **13**, 1975, 1975.
79. Y. Alaki, T. Yoshimoto, M. Imahari and M. Takeuchi, *Kobunshi Kagaku*, **29**, 397, 1972.
80. A. D. H. Clague, J. A. M. Van Broekhoven and L. P. Blaauw, *Macromolecules*, **7**, 348, 1974.
81. W. Hoffman, P. Kuzay and W. Kimmer, *Plaste u. Kaut.*, **21**, 423, 1974.
82. K.-F. Elgert, G. Quack and B. Stützel, *Polymer*, **16**, 154, 1975.
83. P. T. Suman and D. D. Werstler, *J. Polym. Sci.*, *Polym. Chem. Ed.*, **13**, 1963, 1975.
84. H. Sato, A. Ono and Y. Tanaka, *Polymer*, **18**, 580, 1977.
85. Y. Tanaka, H. Sato, A. Ogura and I. Nagoya, *J. Polym. Sci.*, *Polym. Chem. Ed.*, **14**, 73, 1976.
86. Y. Tanaka and H. Sato, *Polymer*, **17**, 413, 1976.
87. M. M. Coleman, D. L. Tabb and E. G. Brame, *Rubber Chem. Tech.*, **50**, 49, 1977.
88. K.-F. Elgert and W. Ritter, *Makromol. Chem.*, **177**, 2021, 1976.
89. Y. Suzuki, Y. Tsuji and Y. Takegami, *Macromolecules*, **11**, 639, 1978.
90. W. Ritter, K.-F. Elgert and H.-J. Cantow, *Makromol. Chem.*, **178**, 557, 1977.
91. C. J. Carman and C. E. Wilkes, *Macromolecules*, **7**, 40, 1974.
92. H. Y. Chen, *J. Polym. Sci.*, *Polym. Lett. Ed.*, **12**, 85, 1974.
93. K. J. Ivin, D. T. Laverty, J. H. O'Donnell, J. J. Rooney and C. D. Stewart, *Makromol. Chem.*, **180**, 1989, 1979.
94. K. J. Ivin, G. Łapiensis and J. J. Rooney, *Polymer*, **21**, 367, 1980.
95. K. J. Ivin, G. Łapiensis and J. J. Rooney, *Polymer*, **21**, 436, 1980.
96. J. P. Kennedy and J. E. Johnston, *Adv. Polym. Sci.*, **19**, 57, 1975.
97. H. R. Kricheldorf and G. Schilling, *Makromol. Chem.*, **179**, 2667, 1978.
98. C. W. Wilson, III, *J. Polym. Sci.*, *A*, **1**, 1305, 1963.
99. J. R. Ebdon, *Polymer*, **19**, 1232, 1978.
100. M. M. Coleman and E. G. Brame, *Rubber Chem. Tech.*, **51**, 668, 1978.
101. A. S. Khatchaturov, E. R. Dolinskaya, L. K. Prozenko, E. L. Abramenko and V. A. Kormer, *Polymer*, **18**, 871, 1977.
102. E. R. Dolinskaya, A. S. Khatchaturov, I. A. Poletayeva and V. A. Kormer, *Makromol. Chem.*, **179**, 409, 1978.

103. T. Asakura, I. Ando, A. Nishioka, Y. Doi and T. Keii, *Makromol. Chem.*, **178**, 791, 1977.
104. A. Zambelli, P. Locatelli and E. Rigamonti, *Macromolecules*, **12**, 156, 1979.
105. Y. Doi, *Macromolecules*, **12**, 248, 1979.
106. T. Fischer, J. B. Kinsinger and C. W. Wilson, III, *J. Polym. Sci.*, *B*, **4**, 379, 1966.
107. F. Keller and M. Arnold, *Plaste u. Kaut.*, **19**, 269, 1972.
108. F. Keller, *Plaste u. Kaut.*, **23**, 730, 1976.
109. E. Klesper, *J. Polym. Sci.*, *B*, **6**, 313, 1968.
110. E. Klesper and W. Gronski, *J. Polym. Sci.*, *B*, **7**, 739, 1969.
111. E. Klesper, A. Johnsen, W. Gronski and F. W. Wehrli, *Makromol. Chem.*, **176**, 1071, 1975.
112. C. J. Carman, R. A. Harrington and C. E. Wilkes, *Macromolecules*, **10**, 536, 1977.
113. J. C. Randall, *Macromolecules*, **11**, 33, 1978.
114. J. R. Ebdon, in *Developments in Polymer Characterisation—2*, J. V. Dawkins (Ed.), Applied Science Publishers Ltd, London, 1980, Ch. 1.
115. M. Delfini, A. L. Segre and F. Conti, *Macromolecules*, **6**, 456, 1973.
116. B. Ibrahim, A. R. Katritzky, A. Smith and D. E. Weiss, *J. Chem. Soc.*, *Perkin Trans.*, **2**, 1537, 1974.
117. T. K. Wu, D. W. Ovenall and G. S. Reddy, *J. Polym. Sci.*, *Polym. Phys. Ed.*, **12**, 901, 1974.
118. W. Hoffman and F. Keller, *Plaste u. Kaut.*, **21**, 359, 1974.
119. F. Keller, *Plaste u. Kaut.*, **22**, 8, 1975.
120. T. K. Wu, *J. Polym. Sci.*, *Polym. Phys. Ed.*, **14**, 343, 1976.
121. T. Moritani and H. Iwasaki, *Macromolecules*, **11**, 1251, 1978.
122. T. K. Wu, *J. Phys. Chem.*, **73**, 1801, 1969.
123. F. Keller and H. Roth, *Plaste u. Kaut.*, **22**, 956, 1975.
124. F. Keller, G. Findeisen, M. Raetzsch and H. Roth, *Plaste u. Kaut.*, **22**, 722, 1975.
125. E. G. Brame, Jr., J. R. Harrell and R. C. Ferguson, *Macromolecules*, **8**, 604, 1975.
126. R. E. Block and H. G. Spencer, *J. Polym. Sci.*, *A2*, **9**, 2247, 1971.
127. J. R. Suggate, *Makromol. Chem.*, **179**, 1219, 1978.
128. J. R. Suggate, *Makromol. Chem.*, **180**, 679, 1979.
129. D. B. Bailey and P. M. Henrichs, *J. Polym. Sci.*, *Polym. Chem. Ed.*, **16**, 3185, 1979.
130. U. Johnsen and K. Kolbe, *Kolloid z. Polym.*, **216/217**, 97, 1967.
131. K. Ishigura, H. Ohashi, Y. Tabata and K. Oshima, *Macromolecules*, **9**, 290, 1976.
132. E. G. Brame, Jr. and F. W. Yeager, *Anal. Chem.*, **48**, 709, 1976.
133. K. Ishigura, Y. Tabata and K. Oshima, *Macromolecules*, **6**, 584, 1973.
134. K.-F. Elgert and B. Stützel, *Polymer*, **16**, 758, 1975.
135. K.-F. Elgert, R. Wicke, B. Stützel and W. Ritter, *Polymer*, **16**, 465, 1975.
136. Y. Inoue, K. Koyama, R. Chûjô and A. Nishioka, *J. Polym. Sci.*, *Polym. Lett. Ed.*, **11**, 55, 1973.
137. R. Roussel and J. C. Galin, *J. Macromol. Sci.- Chem.*, *A*, **11**, 347, 1977.
138. A. R. Katritzky, A. Smith and D. E. Weiss, *J. Chem. Soc.*, *Perkins Trans.*, **2**, 1547, 1974.

139. T. A. Gerken and W. H. Ritchey, *J. Appl. Polym. Sci., Appl. Polym. Symp.*, **34**, 17, 1978.
140. H. J. Harwood and R. C. Chang, *Amer. Chem. Soc., Polymer Preprints*, **12**, 338, 1971.
141. K. Udipi, H. J. Harwood, H. Friebolin and H.-J. Cantow, *Makromol. Chem.*, **164**, 283, 1973.
142. A. Wang, T. Suzuki and H. J. Harwood, *Amer. Chem. Soc., Polymer Preprints*, **16**, 644, 1975.
143. T. Tanabe, H. Koinuma and H. Hirai, *Makromol. Chem.*, **181**, 931, 1980.
144. Y. Mori, A. Ueda, H. Tanzawa, K. Matsuzaki and H. Kobayashi, *Makromol. Chem.*, **176**, 699, 1975.
145. J. Schaefer, *Macromolecules*, **4**, 107, 1971.
146. E. O. Stejskal and J. Schaefer, *Macromolecules*, **7**, 14, 1974.
147. C. Pichot and Q.-T. Pham, *Makromol. Chem.*, **180**, 2359, 1979.
148. T. Moritani and Y. Fujiwara, *Macromolecules*, **10**, 532, 1977.
149. N. W. Johnston and P. W. Kopf, *Macromolecules*, **5**, 87, 1972.
150. G. Sielaff and D. O. Hummel, *Makromol. Chem.*, **175**, 1561, 1974.
151. G. Sielaff and D. O. Hummel, *Makromol. Chem.*, **175**, 1579, 1974.
152. J. R. Ebdon and S. H. Kandil, *J. Macromol. Sci.-Chem.*, *A*, **14**, 409, 1980.
153. V. D. Mochel and B. L. Johnson, *Rubber Chem. Tech.*, **43**, 1138, 1970.
154. A. L. Segre, M. Delfini, F. Conti and A. Boicelli, *Polymer*, **16**, 338, 1975.
155. A. R. Katritzky and D. E. Weiss, *J. Chem. Soc., Perkin Trans.*, **2**, 21, 1975.
156. A. R. Katritzky and D. E. Weiss, *J. Chem. Soc., Perkin Trans.*, **2**, 27, 1975.
157. F. Conti, M. Delfini and A. L. Segre, *Polymer*, **18**, 310, 1977.
158. J. C. Randall, *J. Polym. Sci., Polym. Phys. Ed.*, **15**, 1451, 1977.
159. P. Kuzay and W. Kimmer, *Plaste u. Kaut.*, **18**, 743, 1971.
160. T. Suzuki, Y. Takegami, J. Furukawa, E. Kobayashi and Y. Arai, *Polym. J.*, **4**, 657, 1973.
161. G. A. Lindsay, E. R. Santee, Jr. and H. J. Harwood, *J. Appl. Polym. Sci., Appl. Polym. Symp.*, **25**, 41, 1974.
162. A. R. Katritzky and D. E. Weiss, *J. Chem. Soc., Perkin Trans.*, **2**, 1542, 1974.
163. R. Schmolke, W. Kimmer, P. Kuzay and W. Hufenreuter, *Plaste u. Kaut.*, **18**, 95, 1971.
164. K.-F. Elgert, B. Stützel and I. Forgó, *Polymer*, **16**, 761, 1975.
165. T. Suzuki, Y. Takegami, J. Furukawa and R. Hirai, *J. Polym. Sci.*, *B*, **9**, 931, 1971.
166. C. J. Carman, *Macromolecules*, **7**, 789, 1974.
167. J. C. Bevington and J. R. Ebdon, *Makromol. Chem.*, **153**, 173, 1972.
168. J. R. Ebdon, *Polymer*, **15**, 782, 1974.
169. T. Okada, M. Izuhara and T. Hashimoto, *Polym. J.*, **7**, 1, 1975.
170. A. A. Khan and E. G. Brame, Jr., *J. Polym. Sci., Polym. Phys. Ed.*, **14**, 165, 1976.
171. T. Okada and M. Otsuru, *J. Polym. Sci., Polym. Lett. Ed.*, **14**, 595, 1976.
172. T. Okada and M. Otsuru, *J. Appl. Polym. Sci.*, **23**, 2215, 1979.
173. R. E. Cais, J. H. O'Donnell and F. A. Bovey, *Macromolecules*, **10**, 254, 1977.
174. R. E. Cais and G. J. Stuck, *Polymer*, **19**, 179, 1978.
175. M. Iino, H. H. Thoi, S. Shioya and M. Matsuda, *Macromolecules*, **12**, 160, 1979.

176. R. E. Cais and F. A. Bovey, *Macromolecules*, **10**, 169, 1977.
177. H. R. Kricheldorf and W. E. Hull, *J. Macromol. Sci.-Chem.*, *A*, **11**, 2281, 1977, and earlier papers.
178. H. R. Kricheldorf, *Makromol. Chem.*, **179**, 2133, 1978.
179. H. R. Kricheldorf and W. E. Hull, *Macromolecules*, **13**, 87, 1980, and earlier papers.
180. D. E. Dorman, E. P. Otocka and F. A. Bovey, *Macromolecules*, **5**, 574, 1972.
181. T. Hama, T. Suzuki and K. Kosaka, *Kobunshi Ronbun.*, **32**, 91, 1975.
182. F. A. Bovey, F. C. Schilling, F. L. McCrackin and H. L. Wagner, *Macromolecules*, **9**, 76, 1976.
183. J. C. Randall, *J. Polym. Sci., Polym. Phys. Ed.*, **11**, 275, 1973.
184. M. E. A. Cudby and A. Bunn, *Polymer*, **17**, 345, 1976.
185. D. J. Cutler, P. J. Hendra, M. E. A. Cudby and H. A. Willis, *Polymer*, **18**, 1005, 1977.
186. M. J. Roedel, *J. Amer. Chem. Soc.*, **75**, 6110, 1953.
187. K. B. Abbas, F. A. Bovey and F. C. Schilling, *Makromol. Chem. Suppl.*, **1**, 227, 1975.
188. F. A. Bovey, K. B. Abbas, F. C. Schilling and W. H. Starnes, *Macromolecules*, **8**, 437, 1975.
189. A. Zambelli, P. Locatelli and G. Bajo, *Macromolecules*, **12**, 154, 1979.
190. A. Caraculacu and E. Bezdadea, *J. Polym. Sci., Polym. Chem. Ed.*, **15**, 611, 1977.
191. U. Schwenk, I. Konig, F. Cavagna and B. Wrackmeyer, *Angew. Makromol. Chem.*, **83**, 183, 1979.
192. H.-G. Horn and H. C. Marsmann, *Makromol. Chem.*, **162**, 255, 1972.
193. Y. Mukoyama, T. Tanno, H. Yokokawa and J. Fleming, *J. Polym. Sci., Polym. Chem. Ed.*, **11**, 3193, 1973.
194. A. J. J. De Breet, W. Dankelman, W. G. B. Huysmans and J. De Wit, *Angew. Makromol. Chem.*, **62**, 7, 1977.
195. M. I. Siling, Ya. G. Urman, I. V. Adorova, S. G. Alekseeva, O. S. Matyukhina and I. Ya. Slonim, *Polym. Sci. USSR*, **19**, 358, 1977.
196. J. R. Ebdon and P. E. Heaton, *Polymer*, **18**, 971, 1977.
197. I. Ya. Slonim, S. G. Alekseeva, Ya. G. Urman, B. M. Arshava, B. Ya. Aksel'rod and L. N. Smirnova, *Vysokomol. Soedin.*, *A*, **19**, 793, 1977.
198. I. Ya. Slonim, S. G. Alekseeva, Ya. G. Urman, B. M. Arshava and B. Ya. Aksel'rod, *Vysokomol. Soedin.*, *A*, **20**, 1477, 1978.
199. I. Ya. Slonim, S. G. Alekseeva, Ya. G. Urman, B. M. Arshava, B. Ya. Aksel'rod, I. M. Gurman and L. N. Smirnova, *Vysokomol. Soedin.*, *A*, **20**, 2286, 1978.
200. B. Tomita and S. Hatano, *J. Polym. Sci., Polym. Chem. Ed.*, **16**, 2509, 1978.
201. M. Dawbarn, J. R. Ebdon, S. J. Hewitt, J. E. B. Hunt, I. E. Williams and A. R.Westwood, *Polymer*, **19**, 1309, 1978.
202. H. Schindlbauer and J. Anderer, *Angew. Makromol. Chem.*, **79**, 157, 1979.
203. W. L. Earl and D. L. Vander Hart, *Macromolecules*, **12**, 762, 1979.
204. D. L. Vander Hart, *Macromolecules*, **12**, 1232, 1979.
205. J. Schaefer, S. H. Chin and S. I. Weissman, *Macromolecules*, **5**, 798, 1972.
206. C. A. Fyfe, J. R. Lyerla, W. Volksen and C. S. Yannoni, *Macromolecules*, **12**, 757, 1979.

207. H. A. Resing and W. B. Moniz, *Macromolecules*, **8**, 560, 1978.
208. J. Schaefer, E. O. Stejskal and R. Buchdahl, *Macromolecules*, **10**, 384, 1977.
209. R. H. Atalla, J. C. Gast, D. W. Sindorf, V. J. Bartuska and G. E. Maciel, *J. Amer. Chem. Soc.*, **102**, 3249, 1980.
210. W. L. Earl and D. L. Vander Hart, *J. Amer. Chem. Soc.*, **102**, 3251, 1980.
211. T. K. Wu, *Macromolecules*, **13**, 74, 1980.
212. P. Mansson and L. Westfelt, *Cellul. Chem. Technol.*, **4**, 13, 1980.
213. F. Keller, *Plaste u. Kaut.*, **26**, 136, 1979.
214. R. Lukas, M. Kolinsky and D. Doskocilova, *J. Polym. Sci., Polym. Chem. Ed.*, **17**, 2691, 1979.
215. T. C. Cheng, V. D. Mochel, H. E. Adams and T. F. Longo, *Macromolecules*, **13**, 158, 1980.

Chapter 3

INFRARED AND RAMAN SPECTROSCOPY

W. F. MADDAMS

BP Research Centre, Sunbury-on-Thames, Middlesex, UK

3.1. INTRODUCTION

During the past three decades infrared spectroscopy has become established as a versatile and widely used technique. It came into prominence initially as a major characterisational technique in organic chemistry but it now finds much wider application, for the qualitative and quantitative analysis of materials as diverse as petroleum products, pharmaceuticals, essential oils, pollutants and polymers. Arguably, its most important use is in the last of these areas, where it provides information on topics as varied as fundamental structural studies and the formulation of commercial plastic compositions. In both instances it has been a significant factor in the substantial progress that has been made.

The great majority of the functional groups present in polymers give characteristic bands in the infrared spectral region, thus facilitating their identification. Although the overall structural determination is often an essentially empirical process, requiring comparison of the unknown spectrum with a set of reference spectra it is, nevertheless, very effective. Furthermore, the intensities of the characteristic bands are related to the concentrations of the functional groups in question, thus providing the basis for quantitative analysis. The problem in practice is primarily that of calibration and it often proves advantageous to use the technique in conjunction with elemental analysis and nuclear magnetic resonance spectroscopy.

Numerous qualitative and quantitative applications have been reported. They cover hydrocarbon and vinyl polymers and copolymers, the wide field of ester resins, nylons, polyurethanes, thermosetting resins, rubbers and more specialised materials such as fluoropolymers, polycarbonates, polyethers, polysulphones and cellulose esters. The present account can do no

51

more than give examples selected to illustrate the scope and diversity of the technique. Interested readers may then wish to refer to more detailed texts, such as the excellent 'Identification and Analysis of Plastics', by Haslam *et al.*[1] which contains a useful set of reference spectra chosen as representative of the various types of polymers, and 'Infrared Analysis of Polymers, Resins and Additives: An Atlas' by Hummel.[2] The latter is divided into two parts; the first covers methods for identification in a detailed way and the second is an extensive collection of reference spectra. The volumes of Haslam *et al.*[1] and of Hummel[2] have the additional merit that they discuss infrared spectroscopy in the context of its integrated use with other analytical techniques, where appropriate.

Commercial plastic formulations contain a variety of additives such as plasticisers, fillers, stabilisers and antioxidants. The characterisation of these materials, in which infrared spectroscopy plays an important role, is a complementary subject in its own right and will not be considered here. The topic will be dealt with only so far as is relevant in the context of such materials hindering the identification of the polymers into which they are incorporated.

The value of infrared spectroscopy is considerably enhanced by its ability to provide detailed information on the microstructure of polymers. This includes the way in which successive monomer units add on to the existing chain, both in the case of homo and copolymers, the configurational and conformational structure of the chains and the identification of defect structures and chain end-groups. This type of information is becoming increasingly important in gaining a fuller understanding of the properties of polymers and in improving their long-term stability in commercial usage.

In the decade 1930–1940 infrared spectroscopy played a subordinate role to Raman spectroscopy in the measurement of vibrational spectra. Although the experimental equipment then available lacked sensitivity, and the recording of spectra was very slow and somewhat difficult, the technique was less difficult than its infrared counterpart. During the next thirty years there were tremendous advances in infrared instrumentation, so that modestly priced and reliable spectrometers became widely available. It was only with the advent of the laser source that a renaissance of Raman spectroscopy occurred and its use is still limited by the cost of the equipment. Nevertheless, it does have some fundamental advantages and, so far as polymer studies are concerned, the foremost of these is its ability to handle specimens of varying shapes and sizes. Hence, it is possible to examine materials as they are obtained from processing operations such as extrusion, moulding and blowing, and not have to subject them to the sample preparation procedures used for infrared spectroscopy, which may modify the structural features present as the result of the fabrication operations. To date, Raman spectroscopy has been used primarily to obtain microstructural information but it would be premature to dismiss it as a technique for the qualitative and quantitative characterisation of polymers. It is likely to find increasing use, to complement infrared spectroscopy, in various structural studies on polymers.

3.2. INFRARED SPECTROSCOPY

3.2.1. Experimental
3.2.1.1. The Spectrometer
A detailed account of the construction and operation of infrared spectrometers would be out of place in the present text. Nevertheless, despite the reliability and ease of operation of the various instruments now available, some understanding of their basic features is necessary to achieve optimum results and to avoid pitfalls.

The double beam optical system is now almost universal. The radiation from the source is split into two beams spatially, one of which passes through the sample under examination. The amount of absorption at a particular wavelength is determined by comparing the intensity of the radiation in this sample beam with that of the second beam, the reference channel. Until recently this had usually been done by the null balance principle. The difference in detector signal between the two beams is used to drive an attenuator which is moved into the reference beam until a balance is achieved. This balance point is measured either in terms of percentage absorption or absorbance; the latter is the logarithm of the ratio of the incident to transmitted intensities, which should be proportional to the concentration of the absorbing species. The disadvantage of this photometric system is that at high absorbance values there is little energy in either beam and the system becomes sluggish, making accurate intensity measurements difficult.

Recently, improvements in instrumentation, particularly in detectors, have made possible the direct ratio recording of the sample and reference beam intensities. This is advantageous for quantitative work and it also facilitates measurements on samples which are rather optically dense. Nevertheless, it must be stressed that the best results will be obtained with samples whose thicknesses/concentrations are such that the absorbances of the bands being examined lie in the approximate range 0·1 to 1·0. The relevant sampling methods will be considered in Section 3.2.1.2. The ratio-recording type of photometric system is also advantageous for difference spectroscopy, where one sample is placed in the reference beam, as usual, and the second in the reference beam. This technique is useful for detecting small differences between samples, but it must be used with discretion. For example, it will not reliably detect weak bands that are almost coincident with strong ones.

Bands characteristic for the functional groups commonly encountered in organic compounds lie in the wavelength range 2·5 to 25 μm (4000 to 400 cm^{-1}) and most commercially available instruments are tailored to this requirement. This is readily achievable with a grating monochromator. The energy output of the source varies considerably over this range and this is usually taken into account by programming the width of the spectrometer slits to maintain constant energy. It is necessary to achieve a compromise between the slit width, which partially determines the optical resolution, the scanning speed and the signal to noise ratio of the spectrum, and the manufacturers usually provide a reasonable choice. Typically, a spectrum may be scanned at 2 cm^{-1} resolution, which is adequate for most purposes, with a signal to noise ratio of 500:1, in 5 min.

Analogue output systems are in common use and there is often the facility for abscissal and ordinate expansion, so that selected regions of the spectrum may be examined in more detail. There is also a steady move towards digital recording and this is advantageous when an appreciable number of quantitative measurements of a given type are required. It is also possible to enhance the quality of the spectrometer output by data processing operations such as smoothing and signal accumulation. On-line microcomputers are available as optional accessories for most of the better quality instruments and they are programmed for these data manipulations, together with others such as the calculation of difference spectra.

3.2.1.2. Sample Preparation

In the case of low molecular weight organic compounds four sampling techniques suffice to cope with the majority of materials. Liquids may be examined as thin films, obtained by squeezing a drop between a pair of infrared transmitting plates, or as solutions in solvents such as carbon tetrachloride, cyclohexane, chloroform and carbon disulphide which are reasonably transparent to infrared radiation and have good solvating properties. Solids may also be examined in solution. When the solubility is too low, mulls with liquid paraffin or potassium bromide discs provide alternatives. With the latter technique about 1 % of the sample, as a finely divided powder, is mixed with potassium bromide and cold sintered under pressure to an optically clear material.

These four methods are suitable for the examination of polymers, although more care, experience and sometimes ingenuity are required to obtain satisfactory results. The intensity range of the various absorption bands encountered is such that optimum film thicknesses differ by an order of magnitude. In the case of a saturated hydrocarbon polymer, such as polyethylene, 0·3 mm will suffice whereas with materials containing oxygenated groups, e.g. poly(methyl methacrylate), 0·03 mm is necessary. The latter represents the limit obtainable by hot pressing, a convenient and relatively rapid method that is usually employed with thermoplastic materials. In essence, the procedure is to press the material between polished stainless steel plates, using a temperature at which plastic flow occurs readily, in a small hydraulic press generating a ram pressure of approximately 200 kN. For films with thicknesses greater than about 0·1 mm feeler gauge blades may be used as spacers but for thinner films the sample thickness is adjusted by altering the amount of material used or the pressing temperature.

The element of trial and error in the hot pressing of films is also present in the case of films cast from solution. The primary requirement is for a liquid with a good solvent action but its boiling point is also relevant. If it is too volatile poor quality films are often obtained, whereas a solvent of low volatility will be difficult to remove and may give interfering bands in the spectrum. Artefacts may also occur with solvents, such as tetrahydrofuran, which oxidise on storage. Nevertheless, there is a reasonable choice, among which toluene and 1,2-dichloroethane are probably the most useful. Dimethyl formamide has good solvating power but is rather difficult to remove; water washing of the dried film is probably the best approach. Water can prove a useful solvent for polymers having appreciable concentrations of carboxyl, hydroxyl or amine groups and formic acid is useful for polyamides. Here, also, water washing of the dried film is advisable.

In preparing the solution for casting the upper concentration limit may be set by the need to avoid too viscous a solution; values up to about 5 % are usually suitable. The solution may be placed on a glass plate and the film subsequently stripped off but, if brittleness is a problem, the film may be cast on to a sodium chloride or a potassium bromide window for direct insertion into the spectrometer. An infrared lamp is useful for film drying and final traces of solvent may be removed in a vacuum oven. With some experience it is possible to judge the volume of solution required to obtain a film of a particular thickness. Aqueous lattices may be cast on to silver chloride or KRS5 (thallium bromo-iodide).

It may also be possible to prepare thin films by the use of a microtome. Polymers which are rigid and tough are best suited to this approach and some control in this direction can be obtained by warming or cooling the specimen, as appropriate. Swelling with a solvent prior to sectioning may also prove beneficial. In general, however, the microtome approach is best reserved for the limited number of cases where other methods have failed.

The majority of polymer solvents give strong infrared absorption and examination in solution is therefore not usually a practical proposition. Nevertheless, this approach can be useful in some circumstances and should not be overlooked, particularly for quantitative work. For example, despite its volatility, carbon disulphide is a useful solvent for quantitative work on polybutadiene, butadiene/styrene copolymers and poly-chloroprene.

Polymer samples are not usually available as fine powders and are not readily convertible to this form. Nevertheless, it may be necessary to examine them thus with materials which are insoluble and will not hot press. Various types of rubbers come into this category and these are difficult to comminute. Low temperature grinding is the best approach and suitable equipment is now available commercially. When finely divided specimens are submitted for examination they have usually been recovered from solution, often following a separational procedure such as gel permeation chromatography. Consequently, the amounts available are usually small and there is often insufficient for hot pressing. In these circumstances the use of a mull or a potassium bromide disc may be the only viable approach. The former technique has largely fallen into disuse, for two reasons. The nujol (liquid paraffin) used for the mulling gives its own bands which interfere badly in the case of hydrocarbon polymers. Secondly, although quantitative work on powder samples is not particu-larly easy the pressed disc does have considerable advantages over the mull for this type of work. It is usually possible to obtain satisfactory spectra, in which the background from scattering is not serious, from finely divided polymer specimens sampled as potassium bromide discs.

3.2.1.3. Attenuated Total Reflection Spectroscopy

Attenuated total reflection (ATR) spectroscopy is a technique for obtaining an infrared reflection spectrum which, superficially at least, is very similar to the corresponding absorption spectrum. Although the optical principles involved have been known for many years the technique has only been in use for about two decades. When a beam of radiation is totally internally reflected at the interface between two media there is a small penetration, typically to a depth of a few microns, into the lower refractive index medium. Consequently, if the latter is the sample under investigation it gives a hybrid type of reflection/absorption spectrum, commonly known as an ATR spectrum, which may be used for characterisational purposes. The high refractive index medium must be transparent to infrared radiation and three materials, silver chloride, KRS5 and germanium are in common use. The ATR spectrum approximates more closely to the absorption spectrum as the refractive index difference between the two media increases but the band intensities decrease. Consequently, a compromise is necessary. In view of the small depth of penetration the biggest practical problem is to achieve good optical contact between the sample and the higher refractive index material, which is flat and polished. Hence, the sample needs to be smooth and planar, and to have some flexibility so that it may be gently squeezed against the optical medium. Nevertheless, there are limits to what may be achieved and this has led to the widespread use of multireflection units, which are reasonably effective. Attenuated total reflection spectroscopy is an essentially qualitative technique but it has proved of considerable value and its largest area of application in the polymer field is probably for the examination of surface coatings.

3.2.1.4. Sample Pre-treatment
Many commercial plastic formulations contain a variety of fillers, stabilisers, plasticisers, additives and related materials. These may make the sample opaque to infrared radiation or, less seriously, may give interfering bands. Fillers such as carbon black and titanium dioxide are usually apparent from visual inspection and should be removed prior to spectroscopic examination. Selective extraction of the polymer is often effective, although finely divided carbon black may stay in suspension and prove difficult to separate. Plasticisers are often present at appreciable levels. They may be removed by selective extraction but, unless the polymer is finely divided, their removal may not be complete. Stabilisers and additives are normally present at rather low levels but may cause problems if other low intensity bands are being examined.

In the case of intractable materials not amenable to solvent extraction the examination of pyrolysates may prove useful. In its simplest form the method involves heating the sample under vacuum and condensing the degradation products. This approach leads to non-reproducible results and breakdown products of rather indefinite composition, because of partial repolymerisation. With the advent of the Curie point pyrolyser and low volume multireflection gas cells the preferred approach is to sample directly in the vapour phase or, in the case of complex pyrolysis products, to use prior separation by gas chromatography.

3.2.2. Polymer Identification

3.2.2.1. General Principles
The success of infrared spectroscopy in the characterisation of organic compounds is the result of the almost general validity and applicability of the concept of group frequencies. In a comparatively complex molecule individual functional groups such as carbonyl, hydroxyl, amine and olefine are vibrationally independent of the rest of the molecule and give readily recognisable characteristic vibrational frequencies. There are instances where coupling occurs between vibrational modes, the most notable being amides, but these cases are well documented. The intensity of a particular band is related to the change in dipole moment during the vibration in question and if this change is small, as is the case with the $C{=}C$ stretching mode when this group is near to the centre of molecules, and is also true with C—S and S—S groups, the bands may be too weak to be of diagnostic value. Nevertheless, such limitations are of minority occurrence.

The characterisation of polymers by infrared spectroscopy represents one specific application, albeit a large and important one, of the group frequency concept. To a first approximation the spectrum of a single unit in a homopolymer is identical with that of a long sequence of such units and where differences do occur they are capable of providing valuable information on microstructure, such as stereoregularity and the alternative ways in which successive monomer units add, and also on the way in which the chains pack spatially. With copolymers the spectrum may not be the precise sum of the component bands, because of interaction effects between unlike units. Here also, these effects may provide useful information, such as the diagnosis of random or block copolymers. The literature is replete with examples of the value of infrared spectroscopy for polymer characterisation; in the following section a limited but representative selection of these will be considered.

3.2.2.2. Specific Examples
Figure 3.1 shows what is probably the simplest polymer spectrum. There are three very strong bands at about 2950, 1460 and $720/730 \, \text{cm}^{-1}$. These are assignable to the C—H stretching, deformation and rocking modes of

WAVENUMBERS (cm⁻¹)

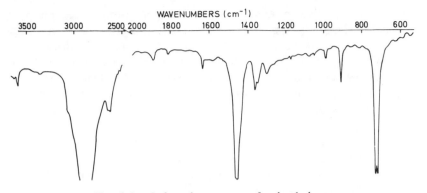

FIG. 3.1.   Infrared spectrum of polyethylene.

CH_2 groups. The absence of other bands of significant intensity precludes the presence of other functional groups. The spectrum is therefore consistent with the structure $(CH_2)_n$, that is, polyethylene. In practice, a certain amount of chain branching is present in polyethylene. The material used to obtain the spectrum shown in Fig. 3.1 is almost linear but if significant branching is present it gives an additional band, at $1378\,cm^{-1}$, overlapping a weakish CH_2 band at $1365\,cm^{-1}$. It is readily observable in low density polyethylene and in ethylene/propylene copolymers.

In the case of polypropylene, which has a methyl branch on every second carbon atom, the CH_3 band at $1378\,cm^{-1}$ is very strong (Fig. 3.2). Additional bands appear, at 970 and $1155\,cm^{-1}$, which are characteristic for $[CH_2CH(CH_3)]_n$ and there is no difficulty in identifying the polymer.

WAVENUMBERS (cm⁻¹)

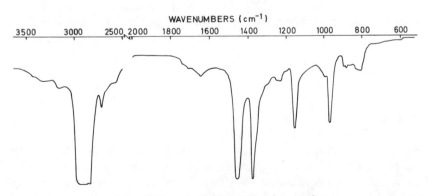

FIG. 3.2.   Infrared spectrum of polypropylene.

With ethylene/propylene copolymers the two types of monomer unit may be disposed randomly or in block fashion. Given that one of them is not present in considerable excess it is possible to distinguish between the two types. With the random polymer there are no long sequences of propylene units and the sole indication for their presence is the strong 1378 cm^{-1} band. However, with the block copolymer there are significant sequences of propylene units and the bands at 970 and 1155 cm^{-1} appear in the spectrum.

Figure 3.3 shows a spectrum which is widely known because its various sharp bands prove useful as wavelength standards for spectrometer calibration. It is particularly rich in bands in the vicinity of 3000 cm^{-1}.

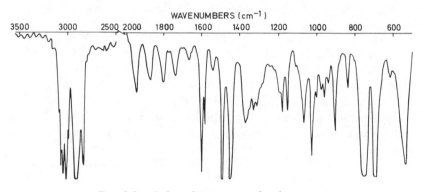

FIG. 3.3.   Infrared spectrum of polystyrene.

Those between 2800 and 3000 cm^{-1} indicate saturated C—H units and those between 3000 and 3100 cm^{-1} =C—H groups. The latter may be present in an olefinic or an aromatic structure but the appearance of the strong band at 1600 cm^{-1}, the aromatic ring 'breathing' vibration, shows that it must be the second of the two possibilities. This is confirmed by the bands at 700 and 760 cm^{-1}, which arise from the out of plane deformation modes of the hydrogen atoms attached to the aromatic ring. Furthermore, their number and positions, together with the bands at 1670, 1740, 1800, 1870 and 1940 cm^{-1}, indicate monosubstituted aromatic rings. There are no additional bands indicative of other functional groups and the only common polymer meeting the above requirements is polystyrene. This deduction is readily confirmed by comparing the spectrum with that of an authentic sample. Although the polystyrene spectrum is rich in band structure it is not difficult to detect when additional peaks are present. A comparison of Figs. 3.3 and 3.4 shows that, in the latter, there is a strong band at 967 cm^{-1}, that the peak at 910 cm^{-1} has intensified relative to the polystyrene peaks at 1030 and 1070 cm^{-1}, and that there is a weak new band at 1640 cm^{-1}. These differences indicate the presence of olefinic C=C units, suggesting that a diene has been copolymerised with styrene. The spectrum is that of a styrene/butadiene copolymer and with this system the butadiene may add 1,2- or *cis*- or *trans*-1,4- addition. The first and last of

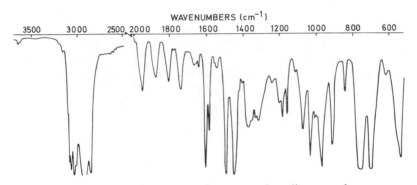

FIG. 3.4. Infrared spectrum of a styrene–butadiene copolymer.

these give the 910 and 967 cm^{-1} peaks. The *cis*-1,4-polymer gives a weak band at about 730 cm^{-1} which is not readily detectable in the presence of an appreciable concentration of polystyrene, and proton NMR spectroscopy provides a more reliable method for characterising materials of this type. The proportions of the 1,2- and *trans*-1,4- units vary with the polymerisation conditions. Hence, the appearance of the spectrum between 900 and 1000 cm^{-1} will be somewhat variable and this must be taken into account when making a characterisation. Terpolymers of styrene, butadiene and acrylonitrile are also commonplace. The presence of the third component is readily deducible by the presence of the characteristic nitrile band at 2240 cm^{-1}.

The spectrum shown in Fig. 3.5 is that of another very familiar polymer. Its general features indicate saturated C—H groups only and bands characteristic for the various types of oxygen-containing groups are absent. Nevertheless, there are some features which are inconsistent with a hydrogen structure. The CH$_2$ deformation band has moved about 30 cm^{-1} to 1430 cm^{-1}, suggesting the presence of a substituent. This is confirmed by

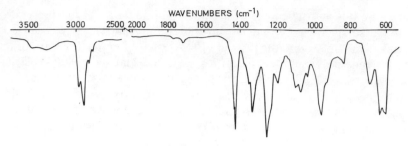

FIG. 3.5.   Infrared spectrum of poly(vinyl chloride).

the bands in the interval 600–700 cm^{-1}, which are C—Cl stretching modes. The two most probable polymers of this type are poly(vinyl chloride) and poly(vinylidene chloride), and comparison with the appropriate reference spectra shows that the unknown is PVC.

Oxygen-containing groups are usually recognisable not only from the wavelengths of the characteristic bands but also by their high intensities. Hence, if they occur in a copolymer, the second component of which is hydrocarbon in type, they stand out even as the minor component. This is the case with the material whose spectrum is given in Fig. 3.6. Comparison with Fig. 3.1 leaves little doubt that polyethylene is one component. The strongest bands, located at 1240 and 1738 cm^{-1}, are indicative of the second component. The second of these peaks is the C=O stretching mode of a carbonyl compound and its precise position suggests a saturated ester. This is confirmed by the 1240 cm^{-1} band, which is still more specific and shows that an acetate group is present. The only likely system of this type is an ethylene/vinyl acetate copolymer and the spectrum matches that of a

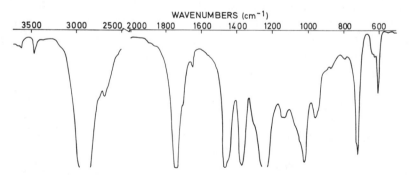

FIG. 3.6.   Infrared spectrum of an ethylene–vinyl acetate copolymer.

reference material containing 15 % w/v of vinyl acetate. The strengths of the two acetate bands, relative to those of the polyethylene, demonstrate very clearly the ability to detect relatively low levels of vinyl acetate. Although PVC has a more intense spectrum than polyethylene, copolymers with vinyl acetate are also readily identifiable by means of the 1240 and 1738 cm^{-1} bands.

Most of the ester groups commonly encountered in polymers give characteristic bands in the interval 1000–1300 cm^{-1}. Figure 3.7 shows a typical example, with two particularly strong peaks at 1130 and 1260 cm^{-1} which are specific for the terephthalate group. Coupled with the sharp band

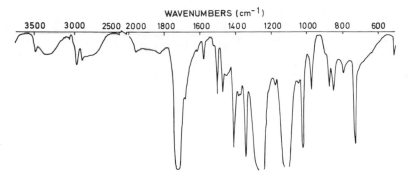

FIG. 3.7.   Infrared spectrum of poly(ethylene terephthalate).

at 730 cm^{-1} this leads to the clear identification of poly(ethylene terephthalate) (terylene). It should be noted that 730 cm^{-1} is an atypical frequency for a p-disubstituted aromatic and this is a consequence of the interaction between the polar ester group and the benzene ring. It is also possible to distinguish between terylene and the other commonly encountered material of this type, poly-bis-1,4-hydroxymethyl terephthalate, by inspection of the region 900 to 1000 cm^{-1}. The spectrum of terylene is sensitive to the degree of crystallinity of the sample and this factor, which will be discussed in more detail in Section 3.2.6, must be taken into account for identification purposes.

The so-called amide I, II and III bands, present in the spectra of compounds containing an amide group, are very characteristic and are readily recognisable. They appear at 1640, 1540 and 1280 cm^{-1}, respectively, in the spectrum of nylon-6,6, shown in Fig. 3.8. Bands at 3070 and 3310 cm^{-1} are also clearly visible; they are the N—H stretching modes

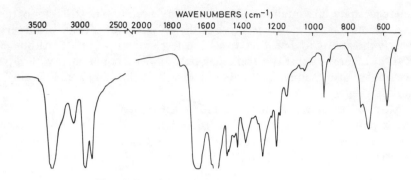

FIG. 3.8.   Infrared spectrum of nylon-6,6.

of the secondary amide group. Having established that a polymer belongs to the nylon group it is possible to go further and to identify it as a particular nylon. This is done by using the weaker bands between 900 and $1300\,\text{cm}^{-1}$, but some care must be exercised because of differences ascribable to the presence of crystalline polymorphs and chains sited in amorphous regions.

Infrared spectroscopy is useful for the characterisation of phenol-formaldehyde resins of the two major types, the resoles prepared in the presence of basic catalysts and the novolaks from acid catalyst condensations. Figure 3.9 shows the spectrum of a typical novolak and it has several characteristic features. The broad band centred at $3350\,\text{cm}^{-1}$ is the hydroxyl stretching vibration of the phenolic group and the peak at $1230\,\text{cm}^{-1}$ is also associated with this functional group. The doublet at $1600\,\text{cm}^{-1}$ is indicative of aromatic rings, as are the peaks at 760 and

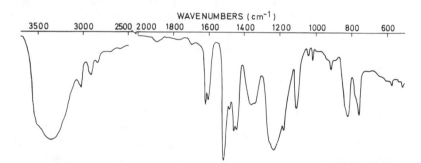

FIG. 3.9.   Infrared spectrum of a novolak type phenol-formaldehyde resin.

$820\,\mathrm{cm}^{-1}$. This pair of bands show that *ortho–para* linkages predominate;

O—O novolaks give a single peak at $755\,\mathrm{cm}^{-1}$.

Uncured resole resins are readily recognisable by a band at $1010\,\mathrm{cm}^{-1}$, arising from methylol groups, although care must be taken to avoid confusion from bands due to hexamine, which is often added as a cross-linking agent. When a resole resin is cross-linked thermally most of the methylol groups disappear and the $1010\,\mathrm{cm}^{-1}$ band is much reduced in intensity. The interested reader may, with advantage, refer to Volume 1, Part 1 of Hummel's 'Infrared Analysis of Polymers, Resins and Additives: An Atlas' for a detailed account of the structural information that may be obtained from the infrared spectra of phenolic resins.

3.2.3. The Quantitative Analysis of Polymers

Quantitative analysis by infrared spectroscopy is based on the application of the Lambert–Beer Law

$$A = \varepsilon c t$$

A is the measured absorbance at the wavelength of the band selected as being characteristic for the component being estimated, ε is the extinction coefficient of the pure component, c is the concentration of the component being estimated and t is the sample thickness. In the case of low molecular weight compounds the procedure is usually straightforward. A solution of known concentration of the sample being examined is placed in a cell of known thickness and the appropriate absorbance value is measured. Furthermore, the pure component is usually available to determine the value of ε.

This sample procedure is sometimes applicable in the case of polymers. In the measurement of the concentration of vinyl acetate in vinyl chloride–vinyl acetate copolymers it is possible to prepare a solution in a solvent such as tetrahydrofuran, and this is examined in a conventional liquid cell. The measured absorbance of the carbonyl band at $1740\,\mathrm{cm}^{-1}$ is compared with the value for pure poly(vinyl acetate), and the composition calculated. This method tacitly assumes that the spectrum of the poly(vinyl acetate) units in the copolymer is the same as that of the homopolymer. This is substantially so in the case of block copolymers but there may be some interaction effects with random copolymers, particularly in view of the fact that the concentration of vinyl chloride units is usually several times that of the vinyl acetate units.

An alternative and sounder approach is to establish a calibration by a more absolute method, such as direct oxygen determination or proton NMR spectroscopy. Once a calibration has been obtained infrared spectroscopy provides a convenient, rapid and cheap method for the analysis of samples on a routine basis. This may be the only feasible approach, particularly when the appropriate homopolymers are not available, or a polymer is being examined to determine the concentration of functional groups formed during degradation or other reactions leading to changes.

Although vinyl chloride–vinyl acetate copolymers are amenable to examination in solution they may also be sampled as thin films; in many other cases this is the only useful approach. Although it is possible, in principle, to measure the thickness of such films, which are often as low as 0·03 mm, they tend to be non-uniform and an average value must be obtained. The need for thickness values is avoided by the use of band intensity ratios. For example, in the case of vinyl chloride–vinyl acetate copolymers it is possible to use the intensity ratio of the acetate band at $1740 \, \text{cm}^{-1}$ to that of the methylene band at $1430 \, \text{cm}^{-1}$. A calibration is obtained from samples of known composition.

The alternative approach is to use much thicker films and to work in the near infrared spectral region. The bands here are overtones and combinations of the fundamental modes in the normal infrared region and are considerably weaker. It is therefore possible to use much thicker specimens and so reduce the measuring errors on this count to negligible proportions. In the case of vinyl chloride–vinyl acetate copolymers a 10 mm thick sheet gives a suitable intensity for the acetate band at $2·15 \, \mu\text{m}$ ($4650 \, \text{cm}^{-1}$). This method is useful for a range of ester copolymers but not all infrared spectrometers are capable of scanning this wavelength region.

In the case of two or multi component systems it is frequently found that there is no wavelength at which one component only absorbs. In these circumstances it is necessary to determine extinction coefficients for each of the components at the wavelengths chosen as most suitable. The resulting set of linear simultaneous equations is then solved to determine the concentrations of the various components. The analysis of styrene–butadiene copolymers for the individual concentrations of *cis-* and *trans*-1,4- and 1,2-polybutadiene, using solutions in carbon disulphide, provides a good example of this approach. It is facilitated by the fact that pure samples of the three types of polybutadiene are available for calibration purposes. When it is necessary to work with films it is still possible to analyse multi component mixtures with overlapping bands using intensity ratios on an empirical basis, *given* that calibrations can be estimated with samples analysed by *another* method. In such cases the plots of intensity ratio against concentration of a particular component may not be linear and may not pass through the origin.

When a straightforward two component system is involved difference spectroscopy may provide the best practical approach. This is exemplified by the determination of chain branching in polyethylene. As noted in the discussion of Fig. 3.1 the methyl groups of chain branches give a band at 1378 cm^{-1}, which is badly overlapped by the stronger methylene band at 1365 cm^{-1}. If a wedge of linear polyethylene or synthetic polymethylene is placed in the reference beam of the spectrometer its position may be adjusted until there is cancellation of the methylene band, leaving a well separated methyl band, as shown in Fig. 3.10. This method should only be used with known systems where there is no possibility of interference from extraneous sources.

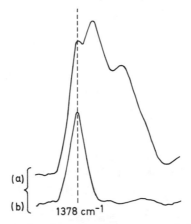

FIG. 3.10. Use of spectral subtraction to measure chain branching in polyethylene. (a) Measured spectrum. (b) Spectrum of the CH_3 band at 1378 cm^{-1} after removal of the overlapping CH_2 absorption.

3.2.4. The Characterisation of Structural Units

The ability of infrared spectroscopy to provide information on the microstructure of polymers has been widely exploited and numerous examples are available. The case of the three types of structural units that occur in polybutadiene has already been mentioned and similar information may be obtained for polyisoprene and polychloroprene where four types of addition, *cis-* and *trans-*1,4-, 1,2- and 3,4- occur. In the case of the simplest polymer, polyethylene, in addition to chain branching there is a second type of defect structure, occasional double bonds. These are readily identifiable from the characteristic out of plane deformation modes. Polymers made by the Phillips process have one end of each chain terminated by a vinyl group, which gives the sharp band at 910 cm^{-1} seen in Fig. 3.1. Low density polyethylenes from high pressure polymerisations contain vinylidene groups, which absorb at 890 cm^{-1}, while polymers prepared with a Ziegler type catalyst frequently have a mixture of three types of unsaturation, the two already mentioned together with trans-CH=CH—, characterised by a band at 965 cm^{-1}.

A sound understanding of the mechanisms whereby polymers degrade thermally and photochemically is of great importance for their commercial usage and infrared spectroscopy has proved very useful for following degradation reactions. The thermal degradation of polyethylene in air proceeds initially via the formation of hydroperoxide groups. These are not particularly easy to detect at low levels but they decompose into a variety of carbonyl-containing compounds whose C=O stretching vibrations are readily visible. The bands overlap but it is usually possible to discern three components, one at 1735 cm^{-1} indicative of saturated aldehydes, the second at 1720 cm^{-1} from saturated ketones and the third, from saturated carboxylic acids, at 1712 cm^{-1}. This third band disappears when the sample is treated with alkali and is replaced by the COO^{-} band at 1610 cm^{-1}, providing useful confirmatory evidence. With photo-oxidised samples these carbonyl bands are again in evidence and are accompanied by peaks at 910 and 990 cm^{-1}, characteristic for vinyl groups, which are probably formed by chain scission. Similarly, infrared spectroscopy proves very useful for the characterisation of the oxidation products of polypropylene.

3.2.5. Configurational and Conformational Isomerism

One major advance in polymerisation chemistry during the last two decades has been the emergence of a range of catalysts for the preparation of stereoregular polymers. For example, with vinyl polymers (—CH$_2$—CHX—)$_n$, some may be obtained in the pure syndiotactic or isotactic forms, that is with the substituent X regularly alternating about the backbone carbons or regularly placed on one side of them. In other cases it has been possible to prepare polymers with a high concentration of one or other of these configurational isomers. Although NMR spectroscopy played the major role in their characterisation many of these configurational isomers have very characteristic infrared spectra. Figure

3.11 gives the spectrum of isotactic polypropylene and comparison with Fig. 3.2 shows that additional sharp bands have appeared at 810, 842, 900, 1000, 1105, 1170, 1260 and 1308 cm^{-1}. Not only are these very characteristic for the isotactic polymer; they may be used to measure the degree of isotacticity. Similarly, the spectrum of isotactic polystyrene has several bands assignable to its stereoregular structure.

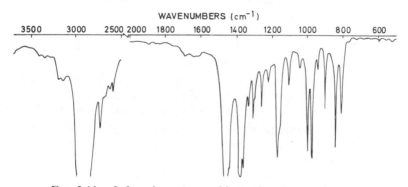

FIG. 3.11.   Infrared spectrum of isotactic polypropylene.

In some cases bands specific for both configurational and conformational isomers are present and poly(vinyl chloride) provides an excellent example. In the interval 600–700 cm^{-1}, where the C—Cl stretching vibrations occur, there are as many as nine overlapping peaks, depending on the polymerisation conditions. They are characteristic for various conformational isomers of the isotactic and syndiotactic configurational isomers. In the case of the former all of the conformers are bent structures, because of steric interference between the chlorine atoms. With the syndiotactic isomer the planar conformer occurs as do higher energy bent structures. The degree of specificity afforded by these bands makes them useful for assessing changes in conformational isomerism consequent upon changes in the polymerisation conditions or as the result of the thermal or mechanical treatment of a sample.

Conformational isomerism also occurs with poly(ethylene terephthalate) and involves the —O—CH$_2$—O— unit. There are two isomers, *gauche* and *trans*; with the former the two oxygen atoms are staggered and with the latter they occur in an extended structure pointing away from each other. Both give several specific bands, of which those at 890 and 1140 cm^{-1} are most characteristic for the *gauche* conformer and those at 840 and 970 cm^{-1} for its *trans* counterpart.

3.2.6. Crystallinity

If there is interaction between adjacent chains in ordered crystalline regions of polymers there may be some changes in the vibrational modes and new bands may appear in the infrared and Raman spectra. There are a number of well documented examples, the best known of which is polyethylene. A doublet is evident at 720 and 730 cm^{-1} in Fig. 3.1. This is the methylene rocking vibration which is split because with this mode the two chains in the orthorhombic unit cell couple symmetrically and anti-symmetrically and the two are not vibrationally equivalent. There has, unfortunately, been a marked failure on the part of various workers to distinguish between bands of this type and those which result from chain regularity. For example, some of the isotactic polypropylene bands shown in Fig. 3.11 have been used as a measure of crystallinity as have the bands at 601 and 638 cm^{-1}, characteristic for syndiotactic sequences, in the spectrum of poly(vinyl chloride). Such use implies that all of these ordered chains are present in crystalline regions but there is a variety of evidence showing that it is not so. Peaks should not be designated as crystallinity bands unless they disappear upon melting, are predicted by group theory and are shown to depend upon the existence of a crystal lattice.

These conditions are not always easy to meet and in practice, the best approach is probably to attempt to correlate the intensities of the suspected crystallinity bands with X-ray diffraction data. For this reason infrared spectroscopy cannot be regarded as a sound method for the diagnosis of crystallinity in polymers. Even when clearly demonstrable crystallinity bands are present their intensities are not necessarily a good quantitative measure of this property. In the case of the 720/730 cm^{-1} doublet of polyethylene the relative intensity is a function of the degree of orientation of the chains in addition to the sample crystallinity. Different crystalline forms of a polymer do not always give different spectra; although the α- and γ- forms of nylon give detectably different spectra, those of the α- and β-forms are virtually identical. Hence, infrared spectroscopy is also an unreliable method for characterising polymorphs.

3.3.   RAMAN SPECTROSCOPY

3.3.1. The Raman Effect

When electromagnetic radiation in the visible region of the spectrum passes through a transparent medium it normally undergoes no change in wavelength. This is because the photons are scattered elastically. However, a very small proportion of them, approximately 1 in 10^6, interact inelastically. They impart energy to the molecules with which they interact and set them into vibrational motion. In order for this to occur there must be a change in polarisability during the vibration in question, in contrast to the infrared absorption process where there must be a change in dipole moment. The inelastically scattered photons lose energy and this is manifest in a shift of the wavelength to the red. For example, if the incident radiation of wavelength 500 nm (20 000 cm^{-1}) excites a vibration at 1000 cm^{-1} the scattered frequency will be 19 000 cm^{-1} and the corresponding wavelength 526·3 nm. The measurement of a shift of 26·3 nm, or even of 2·5 nm corresponding to the excitation of a frequency of 100 cm^{-1}, in the visible spectral region poses no problems. The experimental difficulties relate predominantly to the very low intensity of the scattered radiation. In all except the smallest and most symmetrical molecules the various vibrational modes are permitted in both the infrared and Raman spectra but their intensities often differ markedly. This is sometimes advantageous in that the Raman spectrum of a particular compound may provide the greater sensitivity at low levels. However, the primary advantage, particularly for the characterisation of polymers, is the ease of sampling.

3.3.2. Experimental

The measurement of Raman spectra has been transformed by the advent of the laser. The very low intensity of the Raman scattered radiation calls for a high intensity monochromatic source and the laser meets this requirement admirably. The argon ion laser, together with the less frequently used krypton laser, provides an adequate choice of exciting lines across the visible spectrum. Although it is desirable to use a line in the blue or green spectral region, because of the greater intensity of scattering, the availability of lines in the yellow and red regions is useful for work on coloured samples. The demands on the monochromator used to separate the scattered and incident radiation are considerable, primarily because of their disproportionate intensity ratio. However, the availability of holographic gratings has essentially solved this problem. The performances of the laser and the monochromator, coupled with the efficiency of modern semiconductor detectors, have led to the appearance of commercial instruments which will record Raman spectra almost as rapidly as their infrared counterparts may be obtained. The difference between the two techniques lies in the capital outlay; typically, it is an order of magnitude, in favour of the infrared spectrometer and the differential is unlikely to decrease markedly. This may limit the use of Raman spectroscopy.

The ability to determine vibrational frequencies from measurements in the visible spectral region has marked advantages. Glass containers may be used and, in particular, the tubes may be sealed in the case of volatile or corrosive materials. Water is both transparent to the exciting radiation and is weak in Raman scattering; hence, aqueous solutions are readily amenable to examination, in marked contrast to infrared spectroscopy. The laser source has an additional advantage because the small diameter of the incident beam, which can be as low as about 10 μm, makes it ideal for the examination of small samples. One example is the characterisation of inclusions or 'nibs' in sheet or rod polymer specimens. The limit to the micro sampling capability is usually set by the practical difficulty of locating the exciting beam on the small area to be examined unless a microscope attachment is available.

Sample fluorescence causes an unpredictable and frequently significant problem. Its origin is not fully understood but impurities may often be the source because sample purification frequently reduces it considerably. It is manifest as background emission covering the whole frequency range and in unfavourable cases it can be much more intense than the Raman scattering, and so totally obscure it. Although a change of exciting wavelength may prove beneficial 'burning out' is the most frequent approach. The sample is left in the laser beam, often for several hours, and in many cases the fluorescence gradually diminishes. Although there is insufficient information to draw firm conclusions it does appear that fluorescence may be more troublesome for polymers than for low molecular weight organic compounds and this may limit the use of Raman spectroscopy, particularly for the more routine types of identification work. Fortunately, instrumental developments based on the differing lifetimes of the Raman scattering and fluorescence processes offer hope for a solution to the problem.

3.3.3. Applications
3.3.3.1. Qualitative Aspects
There are advantages and disadvantages in the use of Raman spectroscopy for the characterisation of polymers. Its ability to handle specimens in almost any form means that sample preparation is minimal and, in the absence of fluorescence, a spectrum is probably obtained more rapidly than by infrared spectroscopy. However, in view of the high capital outlay unless a spectrometer is available for other purposes the use of the technique for purely characterisational work, in preference to infrared spectroscopy, is difficult to justify. The most obvious disadvantage is the lack of reference

spectra, although this should be no more than a short-term problem. This is more serious than in the comparable infrared situation because of the very different relative peak intensities and the unfamiliarity of most spectroscopists with such spectra. The point is well illustrated by the Raman spectrum of polyethylene shown in Fig. 3.12. Comparison with the corresponding infrared spectrum (Fig. 3.1) shows that there are comparatively few similarities. The C—H stretching modes appear in both cases and

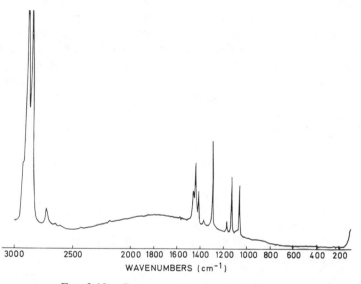

FIG. 3.12. Raman spectrum of polyethylene.

they dominate the Raman spectrum. The CH_2 deformation vibration is also common to both but the well known CH_2 rocking doublet at 720 and $730 \, cm^{-1}$ in the infrared spectrum is absent in the Raman counterpart. This, however, shows characteristic bands at 1070, 1130 and $1300 \, cm^{-1}$, arising from skeletal modes, that is, vibrations of the backbone carbon atoms.

These skeletal modes are a fairly common feature of polymer spectra and are useful for characterisational purposes. For example, they are specific in the case of the various nylons and may be used for the identification of nylons-6, -11, -12, -6,10 and -6,12. Only in the case of nylon-6 and nylon-6,6 are there no useful differences. Sulphur-containing functional groups also give strong and reasonably characteristic Raman spectra. This facilitates the examination of vulcanised samples, for which little information is obtainable by infrared spectroscopy. With the vulcanisation of *cis*-1,4-polybutadiene it has been possible to show the presence of disulphide, polysulphide and five- and six-membered thioalkane and thioalkene units.

The C=C group often shows up more readily in the Raman spectrum, particularly in the case of polymers containing other functional groups which absorb strongly in the region 900 to $1000\,cm^{-1}$ in their infrared spectra, so obscuring the characteristic olefine out of plane deformation modes. The C=C stretching vibration is usually relatively strong in the Raman spectrum and even when it is present at rather low concentrations its position, around $1600-1650\,cm^{-1}$, is such that it is usually not overlapped by other more intense bands. Hence, spectral accumulation, i.e. repetitive scanning into a memory device, may be used to increase its intensity and as previously mentioned this technique, coupled with the high spatial resolution provided by the laser source, make it very useful for characterising small inhomogeneities or 'nibs' in polyethylene. These are usually one of three types of material; high molecular weight polymer, additive particles or extraneous matter such as rust. The second and third types are readily identified and in the case of the Phillips type of polyethylene, where the unsaturation consists of chain terminating vinyl groups, it is possible to identify high molecular weight polymer from the intensity of the C=C stretching band, which decreases with increasing molecular weight.

3.3.3.2. *Quantitiative Aspects*

In infrared spectroscopy the concentration of the absorbing species is proportional to the logarithm of the ratio of the incident and transmitted intensities. Infrared spectrometers are designed to measure intensity ratios and this has the advantage that it eliminates the effect of small fluctuations in the source output, both short- and long-term. With Raman spectroscopy the intensity of a band is linearly related to the concentration of the species from which it originates. Hence, a direct intensity measurement is required and this is difficult on an absolute basis. Consequently, ratio methods are commonly used. With solutions an internal standard may be added; alternatively, one of the solvent bands may prove convenient. With solids it is often possible to find a sample band that may be utilised. For example, in the case of a two component system one band characteristic for each component may be selected.

The determination of polystyrene in nylons is a case in point. The $1003\,\mathrm{cm}^{-1}$ band has been used for the former and the one at $1444\,\mathrm{cm}^{-1}$ for the latter, and the ratio is a linear function of polystyrene content up to the 15% level. Copolymers of vinyl chloride (VC) and vinylidene chloride (VDC), commercially important materials, are readily analysed. In addition to compositional information it is possible to determine the relative amounts of VC—VC, VC—VC—VC, VDC—VDC, VDC—VDC—VDC and VDC—VDC—VDC—VDC units. Professor J. L. Koenig has been the leading worker in both qualitative and quantitative studies during recent years and his account of the Raman scattering of synthetic polymers, in Volume 4 of 'Applied Spectroscopy Reviews', provides additional information and a useful collection of reference spectra.[3]

3.3.3.3. Configurational, Conformational and Other Studies
In many cases Raman spectroscopy provides a valuable method for the study of configurational and conformational isomers. Its specificity is no less than that of infrared spectroscopy and the ease of sampling minimises possible changes in the proportions of the various conformers as no pressing nor grinding is required. For example, although the C—Cl stretching bands in the infrared spectrum are a good indicator for the various conformers of poly(vinyl chloride), it is well established that the proportions of these conformers are changed during sample preparation by techniques such as hot pressing, casting from solution and potassium bromide discs. This problem does not arise in Raman spectroscopy. Furthermore, it is possible to examine relatively large specimens, such as those obtained by extrusion, where the conformational structure may be influenced by the processing operations.

Conformational studies in solution, particularly in the aqueous phase, are also feasible and work on poly(ethylene glycol) is illustrative. When spectra run in water and chloroform solutions are compared with those of the solid and the melt it is evident that the one of the aqueous solution resembles that of the solid polymer, whereas those of the melt and the chloroform solution are similar, and differ from those of the first pair. The detailed deductions, which extend the information obtainable by infrared and NMR spectroscopy, are that the aqueous solution largely retains the TGT conformation of the —C—O—C—O—C— unit that occurs in the solid. In the melt and in chloroform solution the structures are much more conformationally disordered and several isomers are present.

In some cases it is possible to obtain information on longer range structural order, from an examination of the low frequency longitudinal acoustic modes. These are the accordion-like vibrations involving a whole chain segment across the thickness of a lamellar unit such as occurs in polyethylene. The frequencies of these bands, which do not occur in the corresponding infrared spectrum, are inversely proportional to the lamellar thickness and although there are difficulties in using the experimentally determined frequencies to deduce absolute values for thickness the method is very useful on a comparative basis to follow changes such as those occurring during annealing.

3.3.3.4. Resonance Raman Spectroscopy

In most cases, the exciting line used for the Raman scattering is well removed from any electronic absorption bands of the sample being examined. However, this is not the case if the sample is coloured and subject to the further condition that there is interaction between the vibrational and electronic energy levels, there is a very large increase in the polarisability during the molecular vibrations and a corresponding increase in the band intensities. This is the resonance Raman effect, which can lead to intensity enhancements by factors of 10^4 or 10^5. It has not yet proved of interest for polymer studies, with one notable exception, the characterisation of the conjugated unsaturated sequences in degraded poly(vinyl chloride). The mechanism is that initiation occurs at a weak point somewhere along the chain. A molecule of hydrogen chloride is eliminated and this is followed sequentially by further units, in the propagation step, giving conjugated polyene sequences of varying length.

When these sequences are comparatively short, up to eight or nine units, they may be characterised by their ultraviolet/visible absorption spectra. Longer conjugated sequences have their electronic absorption bands in that part of the visible spectrum covered by the various lines of the argon and krypton lasers and give resonance Raman spectra. The method is valuable for three reasons: the $C{=}C$ stretching frequency is dependent on the sequence length, so providing specificity, the sequence lengths that are characterisable begin at the point where ultraviolet/visible spectroscopy loses its specificity, and degradation levels down to $0 \cdot 0001 \%$ are detectable. The method has been used to follow thermal degradation, both in solution and in the solid state, as a function of degradation time and temperature, and the syndiotacticity of the polymer. There are clear indications that very few conjugated sequences of more than about twenty units are present and that termination occurs by cross-linking. The technique has also been used to examine PVC after photochemical degradation and after γ-irradiation.

REFERENCES

1. J. Haslam, H. A. Willis and D. C. M. Squirrell, *Identification and Analysis of Plastics*, 2nd edn., Iliffe, London, 1972.
2. D. O. Hummel, *Infrared Analysis of Polymers, Resins and Additives: An Atlas*, Vol. 1, Part 1, *Text*. Vol. 1, Part 2, *Spectra*. Wiley-Interscience, New York, 1971.
3. J. L. Koenig, *Appl. Spectroscopy Rev.*, **4**, 233, 1971.

BIBLIOGRAPHY

A. Elliot, *Infrared Spectra and Structure of Organic Long-Chain Polymers*, Edward Arnold, London, 1969.

N. J. Harrick, *Internal Reflection Spectroscopy*, Interscience, New York, 1967.

P. J. Hendra, *Adv. Polym. Sci.*, **6**, 151, 1969.

C. J. Henniker, *Infrared Spectroscopy of High Polymers*, Academic Press, New York, 1967.

J. L. Koenig, *Chemical Microstructure of Polymer Chains*, John Wiley and Sons, New York, 1980.

S. Krimm, *Adv. Polym. Sci.*, **2**, 51, 1960.

R. G. J. Miller and B. C. Stace, *Laboratory Methods in Infrared Spectroscopy*, 2nd edn., Heyden and Sons, London, 1972.

H. W. Siesler and K. Holland-Moritz, *Infrared and Raman Spectroscopy of Polymers*, Marcel Dekker, Inc., New York and Basel, 1980.

H. Tadokoro, *Structure of Crystalline Polymers*, John Wiley and Sons, New York, 1979.

R. Zbinden, *Infrared Spectroscopy of High Polymers*, Academic Press, New York, 1964.

Infrared Spectroscopy. Its Use in the Coatings Industry, Federation of Societies for Paint Technology, Philadelphia, 1969.

Chapter 4

ANALYSIS OF POLYMER SYSTEMS BY LUMINESCENCE SPECTROSCOPY

NORMAN S. ALLEN

Department of Chemistry, Manchester Polytechnic, UK

4.1. INTRODUCTION

Over the last twenty years luminescence spectroscopy has proved to be extremely useful as an analytical tool in polymer science. It has provided valuable information on polymer properties such as their macro and supermolecular structure.[1-12] For example, it has been found useful for the determination of the molecular weight of polymers.[9-12] Another, more recent application, however, is its use in the identification of polymer systems[9,10,13,14] and additives in polymers.[15,16] There is a frequent call for the rapid identification of polymer systems, including additives, particularly by scientists and technologists interested in rival products of other companies. In this respect the technique could be useful.

In this chapter a number of representative examples of these uses are highlighted for commercial synthetic polymers together with an account of the theory and origin of luminescence (fluorescence and phosphorescence) emissions and their methods of measurement.

4.2. THEORY OF LUMINESCENCE

When light is absorbed by a molecule, the uptake of energy results in the promotion of an electron from a molecular orbital of lower energy to one of higher energy. The various light emission processes that may occur following the absorption of a quantum of light are best described with reference to a Jablonski diagram.[10,17] A diagram for a simple carbonyl compound is given in Fig. 4.1. The absorption process (A) leads to the formation of excited singlet states (S_1, S_2, etc.). The relative extent of the population of these excited states depends upon the wavelength range of

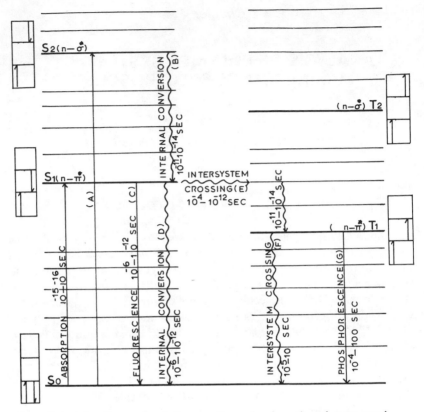

FIG. 4.1.   Energy level diagram for a simple carbonyl compound.

the actinic light. However, as indicated in the diagram, very rapid depopulation of the upper excited states occurs by a process of internal conversion (B) through vibrational relaxation processes.

After deactivation to the first excited singlet state (S_1) several processes are possible for the molecule to reduce its excess electronic energy. It may react chemically, resulting in the formation of a new chemical species or a free radical. The first excited singlet state may lose its excess energy by emitting a photon of light. This emission process is known as fluorescence (C) and generally occurs in about 10^{-8} s, after the formation of the first excited singlet state unless the excess electronic energy can be removed by some other process. It is also possible for the first excited singlet state to lose its energy by some non-radiative process. Thus internal conversion (D), to the ground-state, may occur or, alternatively, since overlap of higher vibrational levels is possible, a non-radiative transition from S_1 to T_1 may occur by electron-spin reorientation even though the process is quantum mechanically forbidden. Such a radiationless transition from a singlet to a triplet state of a molecule is referred to as intersystem crossing (E). Once formed, the triplet molecule rapidly loses to its surroundings any excess vibrational energy and reverts to the lowest vibrational level of the T_1 state.

The molecule remains in this electronically excited state until it can rid itself of the excess energy, either through chemical reaction or by one of the following photophysical processes. The excited molecule in the T_1 state may lose its energy by a non-radiative process of intersystem crossing (F). On the other hand, if the excited molecule in the T_1 state has not returned to the ground-state, $S\sigma$, in a period of time known as the natural lifetime of the excited molecule, it will emit a photon and return to some vibrational level of the ground-state. Again, the excess vibrational energy of the ground-state is rapidly lost to the surroundings. This emission process $(T_1 \rightarrow S_0)$ is referred to as phosphorescence (G) and occurs at longer wavelengths than those at which fluorescence occurs. The natural lifetime of a T_1 state ranges from about 10^{-4} s to several seconds. The fact that the triplet state is much longer than that of the excited singlet state is indicative of the quantum mechanically forbidden nature of the $T_1 \rightarrow S_0$ transition. Hence, a characteristic feature of phosphorescence is an afterglow, i.e. emission which continues after the exciting source is removed. This afterglow is commonly observed from highly crystalline polymers.[9,10,14]

4.3. EMISSION PROPERTIES OF COMMERCIAL SYNTHETIC POLYMERS

Somersall and Guillet[6] have classified polymers into two main types. The first, termed Type A, emit light through isolated impurity chromophores situated as in chain or end-chain groups. The second, termed Type B, emit light through chromophores present in the repeat unit (or units) that form the backbone structure of the polymer. Typical Type A and B polymers may be represented by the general structures (I to VI).

Type A

————————X———————— In-chain chromophore (I)

————————————————X End-chain chromophore (II)

Type B

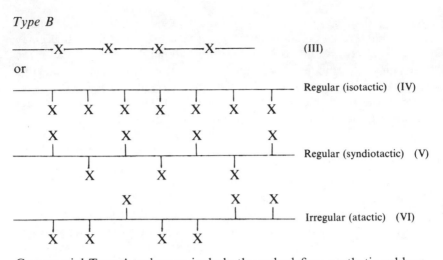

─────•X────•X────•X────•X───────   (III)

or

───────────────────────────── Regular (isotactic) (IV)

X   X   X   X   X   X   X

Regular (syndiotactic) (V)

Irregular (atactic) (VI)

Commercial Type A polymers include the polyolefins, synthetic rubbers, aliphatic polyamides, polyurethanes, poly(vinyl halides), aliphatic polyesters, polystyrenes, polyacrylics and polyacetals. An early study of the luminescence properties of Type A polymers indicated that the fluorescence and phosphorescence emissions may have a common origin irrespective of the chemical structure of the polymer.[18] Recent work with improved spectrofluorimeters, supports this conclusion for a number of polymers although it is not the general case. Typical examples of luminescent chromophores present in Type A polymers are carbonyl and α,β-unsaturated carbonyl groups. These have been found to be responsible for the fluorescence and phosphorescence emissions from polyolefins,[19-21] aliphatic polyamides,[22,23] synthetic rubbers[24] and polyacetals.[25] The latter have the following general structures:

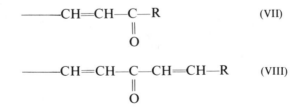

─────CH=CH─C─R   (VII)
              ‖
              O

─────CH=CH─C─CH=CH─R   (VIII)
              ‖
              O

(where R = end of chain or a continuous chain)

─────CH=CH─CH=CH─C─R   (IX)
                    ‖
                    O

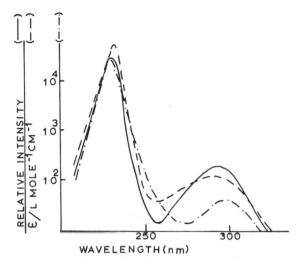

WAVELENGTH(nm)

FIG. 4.2. Comparison of corrected fluorescence excitation spectra of poly(4-methylpent-l-ene) (——) and polypropylene films (–––) with the absorption spectrum of pent-3-ene-2-one (–·–·) in *n*-hexane. (Reproduced from Allen and McKellar, *J. Appl. Polym. Sci.*, **22**, 625, (1978), with permission from John Wiley & Sons, Inc., NY.)

Structure (VII) is an enone (or -al) type and has been found for example, to be responsible for the fluorescence from polyolefins. The evidence for this assignment is shown in Fig. 4.2 where it is seen that the corrected fluorescence excitation spectra of polyolefin films match closely the absorption spectrum of a typical enone compound.[19-21] Structure (VIII) and (IX) are dienones (or -als) and have been found to be responsible for the phosphorescence from polyolefins[19-21] and nylon polymers.[23]

Other commercially important members of the Type A classification are the polyurethanes. A study of an MDI-based polyurethane (diphenyl-methane-4,4'-di-isocyanate) of the general structure (X) has attributed the phosphorescence emission to a benzophenone type chromophore (XI) (Fig. 4.3).[26]

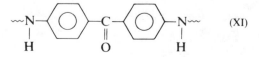

$$\left(ROCO\cdot NH{-}\bigcirc{-}CH_2{-}\bigcirc{-}NHCO\cdot O \right)_n \qquad (X)$$

(where R = aliphatic polyester or glycol residue)

$$\sim\!\!N{-}\bigcirc{-}\underset{\underset{O}{\|}}{C}{-}\bigcirc{-}N\!\!\sim \qquad (XI)$$
$$\;\;\; H \qquad\qquad\qquad\qquad H$$

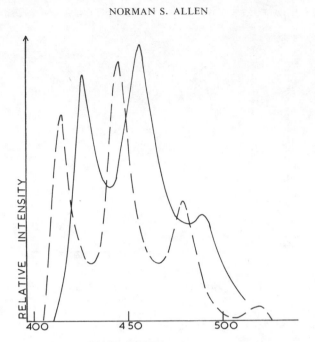

WAVELENGTH, nm

FIG. 4.3. Phosphorescence emission spectra of diphenyl-methane-4,4'-di-isocyanate (——) on MDI-based polyurethane film and (– – –) benzophenone in 2-propanol (77 K). Excitation $\lambda = 325$ nm. (Reproduced from Allen and McKellar, *J. Appl. Polym. Sci.*, **20**, 1441, (1976), with permission from John Wiley & Sons, Inc., NY.)

The luminescence properties of a number of other important Type A polymers are listed in Table 4.1 together with information on the nature of the chromophore, where known.

Commercial Type B polymers include the aromatic polyesters, poly-ethersulphones, aromatic polyamides, some polysulphides, poly-carbonates, poly(phenylene oxides), polyimides and many aromatic resins (e.g. phenolic resins). Apart from the normal fluorescence and phosphorescence emissions from the unit structures of these polymers many of them also exhibit other interesting photophysical phenomena. Owing to the high degree of conjugation in many of these polymers various types of energy transfer processes are possible and these can give valuable information chain conformations. Basically, there are two possible types of energy transfer, namely intermolecular and intramolecular.[6,27]

Intermolecular energy transfer may involve the transfer of energy from

TABLE 4.1
Luminescence Characteristics of Various Commercial Polymers

Polymer		Excitation λ (nm)	Emission λ (nm)	Mean lifetime $\tau_{1/e}$ (s)	Chromophore
		Fluorescence			
Poly(ethylene terephthalate)	Chip	320, 344, 357[a](s)	370, 389, 405	—	Polymer (dimer)
	Film	344, 357	370, 389, 405	—	Polymer (dimer)
	Fibre	344, 357	370, 389, 405	—	Polymer (dimer)
Poly(ethylene-2,6-naphthalate)	Film	375	435	—	Polymer
Polyurethane-MDI[b]-based	Film	372	420	—	Unknown
Nylon-6,6	Chip	357	417	—	α-Keto-imides
	Fibre	357	417	—	α-Keto-imides
Nylon-6	Chip	335	390	—	α-Keto-imides
Nylon-6,10	Chip	345, 355	395, 410	—	α-Keto-imides
Nylon-11	Chip	327, 340	375(s), 395(s), 385	—	α-Keto-imides
Nylon-12	Chip	410	450	—	α-Keto-imides
Poly(vinyl chloride)	Film	290	350	—	Unknown
Poly(tetrafluoroethylene)	Film	328	350	—	Unknown
Poly(vinyl alcohol)	Film	330	360, 370(s)	—	Unknown
Polyethylene (low density)	Powder	258(s), 295, 300	335(s). 350	—	α,β-Unsaturated carbonyl groups of the enone (or-al) type
(high density)	Film	230, 273	295(s). 310, 329(s), 370(s)	—	
Poly(ethylene vinylacetate)	Film	230, 265(s), 290	295(s), 312(s), 330(s), 354(s), 358	—	
Polypropylene	Film	230, 265(s), 290	312(s), 330, 344(s), 358(s)	—	
Poly(4-methylpentene-1)	Film	230, 285	309(s), 320	—	
Polystyrene	Chip	318, 330	336, 354, 368(s)	—	Trans-stilbene
Poly(ethersulphone)	Film	320	360	—	Polymer

TABLE 4.1—contd.

Phosphorescence

Polymer		Excitation λ (nm)	Emission λ (nm)	Mean lifetime $\tau_{1,e}$ (s)	Chromophore
Poly(ethylene terephthalate)	Chip	280, 318, 351	425(s), 360	0·5	Polymer
	Fibre	284, 310	425(s), 477	0·7	Polymer
Poly(ethylene-2,6-naphthalate)	Film	375	580	—	Polymer
Polyurethane-MDI[b]-based	Film	320	423, 489	0·02	Benzophenone type
Nylon-6,6	Chip	296	455, 400	2·10	
	Fibre	296	430	1·30	
Nylon-6	Chip	282	390(s), 420, 455(s)	1·7, 1·6, 1·1	α,β-Unsaturated carbonyl groups of dienone
Nylon-6,10	Chip	300	430	0·70	(or-al type)
Nylon-11	Chip	269(s), 273(s)	423, 450(s)	1·0, 0·88	
Nylon-12	Chip	268, 286(s)	363(s), 410	1·0	
Poly(vinyl chloride)	Film	284	440	0·3	Unknown
Poly(tetrafluoroethylene)	Film	260 – 280[b]	450	0·4	Carbonyl
Poly(vinyl alcohol)	Film	260 – 280[c]	436	0·4	Unknown
Polyethylene (low density)	Powder[d]	273, 280	367, 391, 416	2·30	Benzoic acid
Polyethylene (low density)	Film	278, 280	381, 405, 420	0·60	Benzoic acid
Polyethylene (low density)	Chip[a]	283, 331	370(s), 435, 455	2·15	α,β-Unsaturated carbonyl groups of diene
Poly(ethylene vinyl acetate)	Film	280, 327	455	0·35	(or-al) type
Polypropylene	Film	270, 290, 330	420, 445, 480, 510(s)	0·5–1·2	
Poly(4-methylpentene-1)	Film	273, 330	430	0·86	Unknown
Polyethylene (high density)	Film	275	450	0·35	
Polystyrene	Film	290(s), 300, 336(s)	398, 425, 456, 492	0·008	Acetophenone-end groups
Poly(ether sulphone)	Film	320	450	0·05	Polymer

(s) = Shoulder.
a Prepared using oxygen or aliphatic peroxide.
b Diphenylmethane-4,4'-diisocyanate.
c Broad and structureless spectrum.
d Prepared using a benzoyl-based catalyst.

small molecule chromophores to those on a polymer chain or alternatively from the chromophores on a polymer chain to a small molecule chromophore. Intermolecular energy transfer may also occur between two polymer molecules.

Intramolecular energy transfer on the other hand occurs within the same polymer molecule and in principle can be considered as energy transfer between different parts or segments of the same molecule. There are basically two types of intramolecular energy transfer in polymers. These are (i) between near adjacent conjugated chromophores (XII), and (ii) between non-adjacent or conjugated chromophores (XIII), e.g.

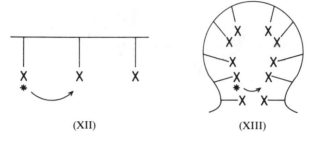

(XII)   (XIII)

In the former process (XII) energy migration (or energy hopping) along the polymer chain can often occur over very long distances. The migrating 'exciton' will move along the polymer chain until it is captured by an impurity or trap (e.g. physical defects in the polymer). When two migrating triplet excitons are in the same polymer chain, mutual annihilation may occur and delayed fluorescence is observed. This emission corresponds with the normal fluorescence spectrum of the polymer but is long-lived since it originates from a triplet state. This process occurs in the poly(vinyl naphthalenes) and poly(vinyl carbazoles).[6]

In the second process (XIII) excimer formation is predominant and has been observed in a number of polymers, e.g. polystyrenes, poly(vinyl carbazoles), poly(vinyl naphthalenes) and poly(vinyl toluenes).[6] An excimer is an excited associated complex formed between a species in the excited singlet state and a similar ground-state species, thus:[6,10,27]

$$A^* + A \longrightarrow (AA^*) \longrightarrow A + A + h\nu \qquad (4.1)$$

Excimer fluorescence occurs to longer wavelengths than the normal fluorescence and is very dependent on the concentration of the species involved. An interesting technique has recently been developed by Phillips and co-workers[28,29] for the study of the time evolution of excited species called time resolved fluorescence spectroscopy. This technique, now actively being used by other workers,[30] has been found to be particularly valuable for the study of, say, excimers in polymer systems. For example, with the use of a time gating system the normal and excimer fluorescence of polymers may be observed separately after excitation.

The fluorescence excitation and emission spectra of a typical Type B polymer, poly(ethylene-2,6-naphthalate), are shown in Fig. 4.4.[31] It is seen that the spectra of the polymer closely match those of a model for the unit structure of the polymer. The fluorescence emission of the polymer, however, is red-shifted from the monomeric emission, indicating the possibility of some excimer emission. The structure of some typical Type B polymers are:

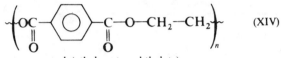

(XIV)

poly(ethylene terephthalate)

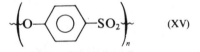

(XV)

poly(ether sulphone)

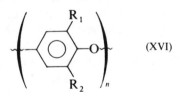

(XVI)

poly(oxy(2,6-dialkyl)-1,4-phenylene)

The fluorescence emission from poly(ethylene terephthalate) has been attributed to an associated ground-state complex formed between the terephthalate units, whereas the phosphorescence emission originates from the monomeric unit structure of the polymer.[32] Emission from an associated ground-state dimer for example occurs by the following reaction:

$$A\text{—}A \xrightarrow{h\nu} (A\text{—}A)^* \longrightarrow A\text{—}A + h\nu \qquad (4.2)$$

In this case both the absorption and emission spectra of the dimer occur at longer wavelengths than the corresponding spectra of the monomeric species.

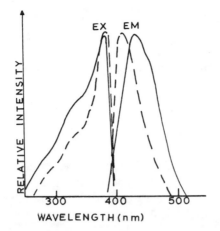

FIG. 4.4.   Comparison of fluorescence excitation and emission spectra (300 K) for poly(ethylene-2,6-naphthalate) film (——) with that of dimethyl-2,6-naphthalate (– – –). (Reproduced from Allen and McKellar, *J. Appl. Polym. Sci.*, **22**, 2085, (1978), with permission from John Wiley & Sons, Inc., NY.)

Finally, another commercially important polymer that has been extensively studied is polystyrene. Evidence from the literature indicates that the fluorescence and phosphorescence spectra of commercial polystyrene are very complex,[10,27,33 – 39] and although features of the fluorescence classify the polymer as one of Type B, the presence of luminescence impurities classify the polymer as Type A.

The fluorescence emission spectrum of polystyrene is often very dependent on the commercial source and purity of the polymer.[37 – 39] All commercial samples of polymer exhibit fluorescence emission at about 310 mm due to residual styrene monomer.[35,39] However, thin films of 'pure' polymer show only excimer emission on excitation at the absorption maximum of the polymer ($\lambda_{max} = 250$ nm).[34,39] In solution, both monomer and excimer emission is observed.[34,36] The intensity of the excimer emission increases with an increase in the concentration of the polymer and a red-shift in the wavelength maximum also occurs.[36] This effect has been attributed to intramolecular quenching of adjacent phenyl chromophores.[35]

Commercial polystyrene also exhibits an 'anomalous' fluorescence with emission maxima at 338, 354, and 872 nm on excitation with light of wavelengths greater than 300 nm.[39] This emission has been attributed to the presence of *trans*-stilbene linkages (XVII).

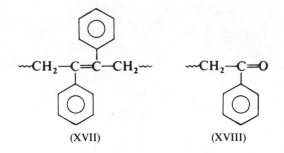

(XVII)   (XVIII)

The phosphorescence emission, on the other hand, has been assigned to the presence of acetophenone-type terminal groups of the structure (XVIII).[37-39] The luminescence properties of a number of other Type B polymers are given in Table 4.1.

4.4. MEASUREMENT OF POLYMER LUMINESCENCE

A typical optical layout for a corrected spectrofluorimeter is shown in Fig. 4.5.[40] Many modern instruments contain two monochromators and two photomultiplier tubes to compensate for variations in the spectral output of the excitation source. The optical layout shown in the figure is for the

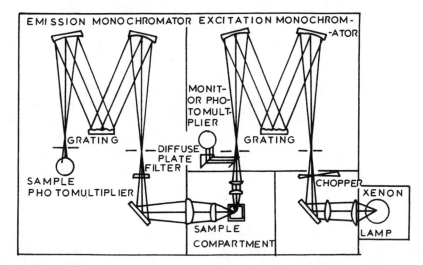

FIG. 4.5.   Typical optical layout for a corrected spectrofluorimeter. (Reproduced from the Hitachi-Perkin-Elmer Manual on the MPF-4 spectrofluorimeter.)

Hitachi-Perkin-Elmer MPF-4 spectrofluorimeter. This instrument employs a 150 W xenon arc lamp as the excitation source and is equipped with two 'red' sensitive R-446F photomultiplier tubes with a wavelength sensitivity of ∼220 to 950 nm. The fluorescence and/or phosphorescence emitted from the sample, after being analysed in the emission monochromator, is detected by the sample photomultiplier amplifier for transmission to the recorder. A portion of the excitation light is led by a beam splitter to the monitor photomultiplier to be converted into an electrical signal. This electrical signal is used for controlling the high voltages that are applied to the 'monitor' and 'sample' photomultipliers in order to compensate for intensity variations in the light source. Spectra are recorded automatically on chart paper.

Phosphorescence lifetime measurements, $>0 \cdot 1$, are obtained by coupling the same intensity signal from the fluorimeter to a storage oscilloscope and storing or photographing the decay of the light spot after chopping the exciting light by means of the excitation shutter. For phosphorescence lifetime measurements, $<0 \cdot 1$ s, the chopper (rotating can) is used as the excitation shutter by adjusting its speed until a stationary waveform of the phosphorescence decay is obtained on the oscilloscope. The mean lifetime, $\tau_{1/e}$, is taken as the time for the phosphorescence intensity to decay to $1/e$ (where $e = 2 \cdot 718$) of its initial value.

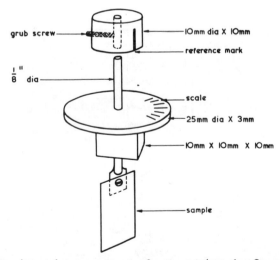

FIG. 4.6. Experimental arrangements for measuring the fluorescence from polymer films. (Reproduced from McKellar and Turner, *Fluorescence News*, 7, 4, (1973).)

Because of their poor solubility many commercial polymers have to be studied in the solid state. Further, because polymers are available in many forms, i.e. powders, granules, fibres, films and sheets, the measurement techniques have to be varied to accommodate the particular sample.

Polymer samples in film form (or sheet) can be easily studied in a fluorescence cell (cuvette) by placing them at an angle of 45° to the beam of excitation light. The film is positioned such that the excitation light is directed away from the emission monochromator. This minimises scattered light from entering the emission monochromator. A modification of the fluorescence technique specifically for studying polymer films has been developed.[41] The experimental set-up is shown in Fig. 4.6 and can be used in many modern spectrofluorimeters. It has many advantages such as good sample intensity reproduction since the angle of the polymer film can be kept constant. Other polymer samples in the form of powders, granules and

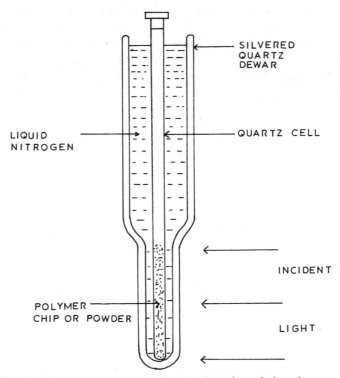

FIG. 4.7.   Experimental arrangement for the detection of phosphorescence from polymer samples.

fibres present problems but can be examined using the phosphorescence attachment. For the measurement of fluorescence however, the rotating chopper has to be removed in order to observe the emission. A silvered Dewar flask like the one shown in Fig. 4.7 can be used as the cell. The powder, granules or fibres can be easily inserted into a silica tube as shown. Larger pieces of polymer chip can be inserted in the cell compartment of the Dewar flask without the tube.

For phosphorescence measurements the Dewar arrangement, shown in Fig. 4.7, is used. In this case, the rotating chopper is put into position, to cut out the fluorescence, and the sample has to be cooled with liquid nitrogen as shown. Film or sheet samples cut into strips can be easily inserted either into the silica sample tube or cell compartment of the Dewar. The Dewar is then turned until maximum emission is observed on the recorder.

4.5. APPLICATIONS

4.5.1. Molecular Structure

Luminescence techniques have been widely used for the study of molecular behaviour in polymers.[1–10,27,42] For example, the familiar glass transition temperature (T_g) is the point at which backbone segmental motion can occur. Above T_g, polymer chain mobility becomes extensive. At temperatures below T_g where chain mobility ceases to occur other transitions are observed due to the rotation of groups either contained in or substituted on the polymer backbone. These molecular motions can affect the phosphorescence yield due to an increase in the rate of non-radiative deactivation processes. A typical plot of phosphorescence intensity against temperature for various ethylene polymers is shown in Fig. 4.8.[42] Two main transitions are observed for these polymers. One at 163 K due to crankshaft-type motion about the polymer chain—the other at 105 K due to rotation of the methyl group adjacent to the carbonyl. Similar results on phosphorescence intensity have been obtained with other polymers and any form of molecular motion tends to increase the probability of conversion of electronic excitation energy to thermal energy, i.e. rotational and vibrational.

Many studies of the behaviour of luminescent additives dissolved in solid polymers have yielded information about the polymer itself. The quenching of the phosphorescence of an aromatic additive by oxygen has formed the basis of a method for the measurement of the movement of the gas through the polymer.[5] The internal rheological behaviour of certain polymers has been investigated through the polarisation of the fluorescence

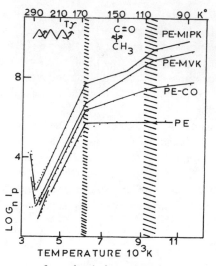

FIG. 4.8.   Arrhenius curves for polyethylene phosphorescence. (Reproduced from Somersall *et al.*, *Macromolecules*, **7**, 233, (1974), with permission from the American Chemical Society.)

of molecules added to or bound to the polymer.[2,3] For example, a clear slightly cross-linked rubber containing auramine O is fluorescent when stretched, but the intensity decreases on relaxation. The luminescence spectra of additives, such as anthracene, in polyethylene are highly sensitive to structural effects in the polymer and vary with the crystallinity of the region in which the emitting molecule is embedded.[43] These studies represent only a few of the many others that have appeared but nevertheless give some idea of the use of luminescence spectroscopy for the study of the molecular structure of polymers.

4.5.2. Molecular Weight

The intensity of the fluorescence and phosphorescence emissions from various synthetic polymers has been related to polymer chain length with varying success.[11,12] For example, for polystyrene the fluorescence intensity of the polymer has been found to decrease with an increase in its molecular weight (Fig. 4.9). From a plot of emission intensity against polymer molecular weight the molecular weight of an unknown sample may be determined.

The luminescence intensity of polyester resins has been found to increase with an increase in the degree of cross-linking.[44] In one specific case, the

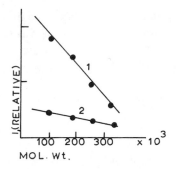

FIG. 4.9. Decrease in intensity of fluorescence of polystyrene at 423 nm(1) and 600 nm(2) with increase in molecular weight. (Reproduced from Gachkovskii, *Polym. Sci.*, (*USSR*), 2199, (1956), original source, *Vysokomol. Soedin.*, A7, 2009, (1965), with permission from Pergamon Press Ltd.)

fluorescence spectra were found to reflect intrachromophore hydrogen bonding as opposed to hydrogen bonding involving chain cyclisation.[45]

4.5.3. Identification of Commercial Polymers

The luminescence properties of polymers can be useful for their identification, particularly if they are used in conjugation with other techniques such as infrared spectroscopy.[9,10,14] The use of luminescence spectroscopy as an analytical technique for polymer identification involves the measurement of the following properties.[14]

(1) The fluorescence emission spectrum, which is obtained by exciting the polymer with light of any wavelength that is capable of producing the emission spectrum. The most satisfactory spectrum for definitive purposes, however, is normally recorded at the wavelength maximum (λ_{max}) of the excitation spectrum.

(2) The fluorescence excitation spectrum which is obtained by recording the variation in intensity of the emission spectrum (λ_{max}) as a function of the excitation wavelength. The corrected excitation spectrum should closely match the absorption spectrum of the chromophore.

(3) The phosphorescence emission spectrum which is obtained by using the same method as used for the fluorescence emission spectrum.

(4) The phosphorescence excitation spectrum which is obtained by using the same method as used for the fluorescence excitation spectrum.

(5) The phosphorescence lifetime which is determined as described above.

Finally, if it is found that the wavelength maximum (λ_{max}) of either the fluorescence or phosphorescence spectrum varies with the excitation wavelength, the presence of more than one chromophore is indicated and it is necessary to examine the polymer over a wide range of excitation wavelengths in order to achieve its complete identification. The above luminescence characteristics of a number of commercial polymers are shown in Table 4.1. The identities of the chromophores that are believed to be responsible for the various emissions are also shown.

All the polymers shown in Table 4.1 exhibit fluorescence emission in the wavelength range 300–450 nm and phosphorescence emission in the range 400–600 nm. In certain cases the spectra are highly structured and this assists further in the identification of the polymer. The highly structured phosphorescence emission spectra of poly(ethylene terephthalate), wool, polystyrene and polyethylene shown in Fig. 4.10 are good examples to illustrate this point. The manufacturing history of a polymer can in some cases influence the nature of the luminescence. This effect is demonstrated in Table 4.1 for polyethylene, where it is seen that the phosphorescence emission of polymer prepared using a benzoyl-peroxide catalyst differs from that of polymer prepared using oxygen as a catalyst.[20]

The advantages and disadvantages of luminescence spectroscopy for identifying polymers are:

Advantages

(1) The technique is rapid and non-destructive.

(2) Its use requires no tedious sample preparation.

(3) It can be used to analyse polymers in powder, chip, film or fibre form. Normally the amounts of polymer required are in the range 10–50 mg for powders and film and up to 100 mg for fibres and chip.

(4) It can be a highly sensitive technique—concentrations of 10^{-8} moles litre^{-1} of aromatic hydrocarbons have been detected by fluorescence spectroscopy.[33]

(5) It can be used to characterise polymers within a particular class, e.g. the polyamides (see Table 4.1). This advantage would be useful in conjunction with other techniques such as infrared spectroscopy.

Disadvantages

(1) The technique gives no information on those commercial polymers which are non-luminescent or only weakly so, e.g. polyacrylics. However, this problem can sometimes be overcome by thermally

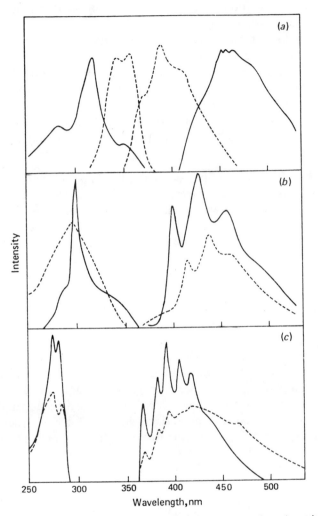

FIG. 4.10.   (a) Fluorescence (– – –) and phosphorescence (——) excitation and emission spectra of poly(ethylene terephthalate) chip; (b) phosphorescence excitation and emission spectra of polystyrene chip (——) and wool fibre (– – –), and (c) phosphorescence excitation and emission spectra of polyethylene powder (——) and film (– – –). (Reproduced from Allen *et al.*, *The Analyst*, **101**, 260, (1976), with permission from the Chemical Society, London.)

oxidising the polymer under controlled conditions. The non-volatile oxidation products may be luminescent and characteristic for that particular polymer.[46]

(2) Some light stabilisers reduce the intensity of the luminescence from polymers[47] and if their presence is suspected, they must be removed by solvent extraction. Similarly, other processing additives such as anti-oxidants may also be present in the extract, some of which may be identified by luminescence spectroscopy.[16]

(3) The presence of certain pigments may prove difficult but they themselves often exhibit their own characteristic emission and can be analysed (see below).

Finally, since the number of polymers listed in Table 4.1 represents only a fraction of those available commercially and because different instruments will give slightly different wavelength values, it is recommended that the analyst constructs a table of data (or spectra) by using his own instrument. When constructing a table it is also recommended that the form of the polymer be noted as this will help to eliminate any differences due to processing history and morphology.

4.5.4. Analysis of Additives in Polymer

Many anti-oxidants and ultraviolet stabilisers exhibit their own characteristic fluorescence and/or phosphorescence emissions and may therefore be analysed after solvent extraction from the polymer. A number of the more recent hindered piperidine light stabilisers however, do not luminesce. The fluorescence and phosphorescence properties of a number of anti-oxidants and ultraviolet stabilisers are shown in Table 4.2. The detection limits of the additives vary quite markedly and this could be a problem particularly for commercial systems containing mixtures of additives. In most commercial polymers an anti-oxidant and ultraviolet stabiliser combination is used and these may have to be separated using TLC if their emission spectra cannot be resolved.

4.5.5. Analysis of Pigments in Polymers

Certain types of pigments (like certain types of dyes) exhibit their own characteristic luminescence spectra. Of the many types of pigments available, white pigments are the most widely used, particularly titanium dioxide.[10,27] The two crystalline modifications of titanium dioxide, anatase and rutile, may be easily identified by their characteristic emissions.[48-51] At low temperatures (77 K) anatase exhibits a strong emission

TABLE 4.2
Luminescence Characteristics of Compounds Examined

No.	Trade name	Chemical composition	Room temp. fluorescence ex λmax	em λmax	Low temp. luminescence ex λmax	em λmax	Phosphorescence ex λmax	em λmax	Phosphorescence life-time (s)	Phosphorescence detection limit (ppm)
1.	Topanol A	2,4-Dimethyl-6-t-butylphenol	255	318[a]	289	425	285	420	0·5	1·0
2.	Topanol OC	2,6-Di-t-butyl-4-methylphenol	—	—	—	—	—	—	—	—
3.	Tenox BHA	Mixture of 2- and 3-t-butyl-4-hydroxyanisole	312, 255	380, 335	299	420	295	420	1·80	0·06
4.	Binox M	Bis-3,5-di-t-butyl-4-hydroxyphenyl-methane	—	—	—	—	—	—	—	—
5.	Ionox 330	1,3,5-Trimethyl-2,4,6-tris-(3,5-di-t-butyl-4-hydroxy-benzyl)benzene	295	335[a]	—	—	—	—	—	—
6.	Nonox WSP	Bis(2-hydroxy-3-α-methyl-cyclohexyl-5-methylphenyl)-methane	372	440[a]	290	415	290	412	1·53	0·1
7.	Nonox WSL	2,4-Dimethyl-6-α-methyl-cyclohexylphenol	372	464[a]	296	430	282	415	0·5	1·0
8.	Nonox DCP	2,2-Bis(3-methyl-4-hydroxyphenyl)propane	310, 370	390, 460[a]	290	410	285	405	1·9	0·05
9.	Calco 2246	Bis(2-hydroxy-3-t-butyl-5-methylphenyl)methane	325	375[a]	282	410	285	410	1·56	0·2
10.	Topanol CA	1,1,3-Tris(2-methyl-4-hydroxy-5-t-butylphenyl)-methane	255, 318	318, 408[a]	280	405	285	405	0·70	0·07
11.	Santonox R	Bis(2-methyl-4-hydroxy-5-t-butylphenyl)sulphide	360	410[a]	300	430	305	430	0·035[c]	0·07
12.	Topanol TP	Bis(2-hydroxy-3,5-di-t-butyl-6-methylphenyl)-sulphide	—	—	295	428	300	426	0·033[c]	0·1
13.	Suconox 18	N-Stearoyl-p-aminophenol	318, 345	386[a]	300	415	300	415	1·0	0·12
14.	Naugawhite	Bis(2-hydroxy-3-nonyl-5-methylphenyl)methane	366, 312	446, 390[a]	285	408	285	410	1·60	0·08
15.	Agerite Superlite	—	345	405	290	410	290	420	1·85	0·06
16.	Voidox 100%	2,6-Di-t-butyl-4-methylphenol + sorbitan/fatty acid compound	266	318[a]	295	420	292	435[b]	2·5	>10
17.	Irganox 1010	Pentaerythritol-tetra-β-(3,5-di-t-butyl-4-hydroxyphenyl)propionate	300	350[a]	328	380	—	—	—	—
18.	Irganox 1076	n-Octadecyl-β-(3,5-di-t-butyl-4-hydroxyphenyl)-propionate	376	430	—	—	—	—	—	—
19.	Irganox 1093	Di-n-octadecyl-3,5-di-t-butyl-4-hydroxybenzyl phosphonate	315	375[a]	—	—	—	—	—	—
20.	Polygard	Tris(nonylphenyl)-phosphite	356	422[a]	280	400	282	400	1·75	0·1
21.	Nonox CI	N,N'-Di-β-naphthyl-p-phenylenediamine	392	490	386, 434	430, 455	382	516	0·90	0·02
22.	DLTDP	Dilaurylthiodipropionate	—	—	—	—	—	—	—	—
23.	Salol	Phenylsalicylate	340	464[a]	320	452	302	415	0·45	0·1
24.	Cyasorb UV9	2-Hydroxy-4-methoxy-benzophenone	—	—	300	450	300	450[b]	0·02	>10
25.	Cyasorb UV531	2-Hydroxy-4-n-octoxybenzophenone	—	—	305	450	300	450[b]	0·03	>10
26.	Uvinol 400	2,4-Dihydroxybenzophenone	—	—	—	—	—	—	—	—
27.	Cyasorb UV24	2,2'-Dihydroxy-4-methoxybenzophenone	—	—	305	430	308/360	430/455[b]	0·1	>10
28.	Tinuvin P	2-(2'-Hydroxy-5'-methylphenyl)-benzotriazole	—	—	315	396, 480	315	480, 515	0·85	0·03
29.	Tinuvin 326	2-(2'-Hydroxy-5'-t-butylphenyl)-3-chlorobenzotriazole	—	—	—	—	—	—	—	—

[a] Denotes weak room temperature fluorescence.

[b] Denotes weak phosphorescence at 77 K. only approximate estimates of life-time made.

[c] Life-time measurement reproducible within ± 10%

Reproduced from Kirkbright et al.[16] with permission of Elsevier Scientific Publishing Co.

spectrum at 540 nm ($\tau_{1/e} = 3$ ms) while rutile exhibits weak emission in the infrared at 815 and 1015 nm ($\tau_{1/e} \simeq 10^{-5}$). The excitation of λ_{max} of the two crystalline forms are 340 and 375 nm, respectively. This difference in emission properties of the two forms has also been associated with their marked difference in photoactivity in a polymer.[48,51] The nature of the surface treatment, often applied to titanium dioxide pigments for various commercial reasons, also affects the intensity of the emissions from the pigments and their photoactivity.[49] The manufacturing history of titanium dioxide pigments can also be determined from their characteristic emission spectra in the infrared. For example, at low temperatures (77 K) rutile pigments manufactured by the 'sulphate' process exhibit emission at 815 and 1015 nm, whereas those manufactured by the 'chloride' process exhibit emission at 1015 nm (Fig. 4.11).[50] Anatase pigments also give an emission

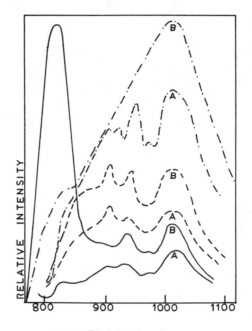

WAVELENGTH (nm)

FIG. 4.11. Emission spectra of a 'sulphate' processed rutile pigment (——) (S × 100), a 'chloride' processed rutile pigment (– –) (S × 100) (A), and a 'sulphate' processed anatase pigment (– · –) (S × 100) (A), (S × 30) (B) in polyethylene at room temperature (300 K) (A) and liquid nitrogen temperature (77 K) (B), respectively. Excitation $\lambda = 375$ nm. S = sample sensitivity.

at 1015 nm but its intensity is much stronger than that from any of the rutile forms.

The luminescence properties of coloured pigments have not been reported on.

REFERENCES

1. S. Czarnecki and M. Kryszenski, *J. Polym. Sci.*, *Polym. Chem. Ed.*, **1**, 3067, 1963.
2. G. Oster and V. Nishijima, *Fortschr. Hochpolym. Forsch.*, **3**, 313, 1964.
3. G. Oster and V. Nishijima, *Newer Methods of Polymer Characterisation*, Wiley Interscience, New York, 1964, Ch. 5.
4. E. I. Hormats and F. C. Underleitner, *J. Phys. Chem.*, **69**, 487, 1968.
5. G. Shaw, *Trans. Farad. Soc.*, **63**, 2181, 1967.
6. A. C. Somersall and J. E. Guillet, *J. Macromol. Sci.*, *Revs. Makromol. Chem.*, **C13**, 135, 1975.
7. J. E. Guillet, *Proceedings of the International Symposium on Macromolecules*, Rio de Janeiro, July 26–31, 1974, p. 183.
8. A. M. North, *Brit. Polym. J.*, **7**, 119, 1975.
9. N. S. Allen and J. F. McKellar, *Chem. & Ind.*, (*London*), 907, 1978.
10. J. F. McKellar and N. S. Allen, *Photochemistry of Man-Made Polymers*, Applied Science Publishers Ltd, London, 1979, Ch. 7.
11. V. F. Gachkovskii, *Vysokomol. Soedin*, **17**, 2009, 1975.
12. V. F. Gachkovskii, *Adv. Mol. Relax. Proc.*, **5**, 24, 1973.
13. H. F. Smith, in *Proceedings of the Symposium on Interdisciplinary Approaches on Polymer Characterisation*, C. P. Craven (Ed.), Plenum Press, New York, 1971, p. 249.
14. N. S. Allen, J. Homer and J. F. McKellar, *The Analyst*, **101**, 260, 1976.
15. H. V. Drushel and A. L. Sommers, *Anal. Chem.*, **36**, 836, 1964.
16. G. F. Kirkbright, R. Narayanaswarmy and T. S. West, *Anal. Chim. Acta.*, **52**, 237, 1970.
17. J. G. Calvert and J. N. Pitts, *Photochemistry*, John Wiley & Sons, Interscience, New York, 1965.
18. V. F. Gachkovskii, *Zhurnal. Strukturnol Khimii*, **4**, 424, 1963.
19. N. S. Allen, J. Homer and J. F. McKellar, *J. Appl. Polym. Sci.*, **21**, 2261, 1977.
20. N. S. Allen, J. Homer and J. F. McKellar, *J. Appl. Polym. Sci.*, **21**, 3147, 1977.
21. N. S. Allen and J. F. McKellar, *J. Appl. Polym. Sci.*, **22**, 625, 1978.
22. H. D. Scharf, C. G. Dievis and H. Leismann, *Angew Makromol. Chemie.*, **79**, 193, 1979.
23. N. S. Allen, J. F. McKellar and D. Wilson, *J. Photochemistry*, **6**, 337, 1976.
24. S. W. Beavan and D. Phillips, *J. Photochemistry*, **3**, 349, 1974.
25. N. S. Allen and J. F. McKellar, *Polym. Deg. & Stab.*, **1**, 47, 1979.
26. N. S. Allen and J. F. McKellar, *J. Appl. Polym. Sci.*, **20**, 1441, 1976.
27. B. Ranby and J. F. Rabck, *Photodegradation, Photooxidation and Photo-stabilisation of Polymers*, Wiley Interscience, New York, 1975.

28. M. D. Swords and D. Phillips, *Chem. Phys. Letts.*, **43**, 228, 1976.
29. K. P. Ghiggino, R. D. Wright and D. Phillips, *J. Polym. Sci., Polym. Phys. Ed.*, **16**, 1499, 1978.
30. C. E. Hoyle, T. L. Nemzek, A. Mar and J. E. Guillet, *Macromolecules*, **11**, 429, 1978.
31. N. S. Allen and J. F. McKellar, *J. Appl. Polym. Sci.*, **22**, 2085, 1978.
32. N. S. Allen and J. F. McKellar, *Makromol. Chemie*, **179**, 523, 1978.
33. A. Charlesby and R. H. Partridge, *Proc. Roy. Soc.*, **A283**, 312, 1965; *idem-ibid*, 329.
34. M. T. Vala, R. Silbey, S. A. Rice and J. Jortner, *J. Chem. Phys.*, **41**, 2146, 1964.
35. L. J. Basille, *J. Chem. Phys.*, **36**, 2204, 1962.
36. T. Nishihara and M. Kaneko, *Makromol. Chemie*, **84**, 124, 1969.
37. G. A. George, *J. Appl. Polym. Sci.*, **18**, 419, 1974.
38. G. A. George and D. K. C. Hodgeman, *Eur. Polym. J.*, **13**, 63, 1977.
39. W. Klopffer, *Eur. Polym. J.*, **11**, 203, 1975.
40. *Manual for the MPF-4 Spectrofluorimeter*, Hitachi-Perkin Elmer.
41. J. F. McKellar and P. H. Turner, *Fluorescence News*, 714, 1973.
42. A. C. Somersall, E. Dan and J. E. Guillet, *Macromolecules*, **7**, 233, 1974.
43. G. P. Egorov and E. G. Moisya, *J. Polym. Sci., Part C*, **16**, 2031, 1967.
44. V. F. Gachkovskii, *Zh. Strukt. Khim.*, **9**, 1018, 1968.
45. V. F. Gachkovskii, *Zh. Strukt. Khim.*, **8**, 362, 1967.
46. N. S. Allen, J. Homer, J. F. McKellar and G. O. Phillips, **7**, 11, 1975.
47. N. S. Allen, J. Homer and J. F. McKellar, *Makromol. Chemie*, **17**, 1575, 1978.
48. N. S. Allen, J. F. McKellar, G. O. Phillips and D. G. M. Wood, *J. Polym. Sci., Polym. Lett. Ed.*, **12**, 241, 1974.
49. N. S. Allen, D. J. Bullen and J. F. McKellar, *Chem & Ind.*, (*London*), 797, 1977.
50. N. S. Allen, D. J. Bullen and J. F. McKellar, *Chem. & Ind.*, (*London*), 629, 1978.
51. N. S. Allen, J. F. McKellar and D. Wilson, *J. Photochemistry*, **7**, 319, 1977.

Chapter 5

ANALYSIS OF POLYMERS BY MASS SPECTROMETRY

SALVATORE FOTI and GIORGIO MONTAUDO
*Istituto Dipartmentale di Chimica e Chimica Industriale,
Università di Catania, Italy*

5.1. INTRODUCTION

Although mass spectrometry (MS) is considered an essential technique for elucidating the structure of low molecular weight organic and inorganic compounds it has been much less used in the case of polymers.

This constitutes a relevant difference with respect to other widely applied spectrometric techniques, such as i.r. and NMR, whose respective importance in the structure elucidation of polymers is about the same as in the field of low molecular weight compounds.

It appears evident that, since MS techniques require transfer of the sample in the gas phase, the low volatility of macromolecules has constituted a serious drawback for the application of mass spectrometric analysis to polymer systems.

Most applications of MS in the field of polymer analysis have been performed interfacing a pyrolytic unit with a GG–MS apparatus. Such a combination is commonly applied to identify the volatile products formed by pyrolysis of polymer samples; for a review of the subject see ref. 1.

Polymer pyrolysis performed externally to the mass spectrometer, followed by direct insertion of the products into the MS, has been in use since the early 1950s,[2-4] but the obvious drawbacks of the external pyrolysis method were apparent.[5]

In these procedures, the mass spectrometer plays its conventional role and therefore will not be discussed further.

Remarkable progress was achieved when polymers were introduced directly in to the mass spectrometer and the pyrolysis was performed close to the ion source.[6-9]

Indeed, it became possible to detect oligomers of considerable size and the increased potentialities of the method became more widely appreciated.[10,11]

Direct MS analysis of polymers started to be applied systematically from about 1965. Since then, nearly 100 original papers have appeared in the chemical literature, covering almost all classes of polymers and copolymer systems. Various aspects of this work have been reviewed by Hummel,[12] Sedgwick,[13] Lüderwald[14] and Wiley.[15]

In spite of this, even now mass spectrometry of polymers remains confined to a few laboratories.

It is the aim of this review to provide a critical discussion of the method and to examine current trends in MS of polymers.

Since polymers are amenable to MS only after thermal degradation, their mass spectral characterisation assumes chemical and structural significance only in the light of the thermal decomposition processes occurring in the pyrolytic stage.

Mass spectroscopic analysis of polymers, beside being useful in polymer and copolymer structural identification, is also a most promising technique for investigating the thermal decomposition mechanisms of polymers.

Therefore, an attempt will also be made to place this technique in the right perspective with respect to conventional thermoanalytical methods.

No special efforts will be made to deal thoroughly with the theoretical background of MS, since references to general aspects of mass spectrometry are readily available.

5.2.   PRINCIPLES OF THE METHOD

5.2.1.   Mass Spectrometric Techniques
In dealing with direct MS analysis of polymers several instrumental variables need to be considered; methods of sample introduction and volatilisation, methods of ionisation, mass separation and mass detection. Furthermore, attention has to be paid to techniques of compounds identification in the mixture of thermal degradation products.

5.2.1.1. Sample Introduction and Volatilisation
Polymer samples are generally introduced in the ion source by the conventional direct inlet for solid samples; Field Desorption (FD) is the only exception (see below).

Since polymers are, in general, not volatile as such, they must be pyrolysed in order to obtain volatile oligomers suitable for MS analysis.

This implies that the mass spectrum of a polymer is actually the mass spectrum of the mixture of oligomers formed by pyrolysis.

There are two general ways in which thermal degradations into the mass spectrometer have been carried out: (1) a rapid heating to a pre-determined temperature where isothermal pyrolysis should take place (Flash Pyrolysis: FPy–MS); or (2) a programmed temperature rise (Gradual Pyrolysis: GPy–MS).

FPy–MS has been performed by Curie-point[16] or laser-heating.[17] The mixture of volatile fragments formed is either introduced directly in the ions source[16,17] or, prior to that, collected in a heated expansion chamber.[18] A disadvantage of FPy–MS is that the thermal degradation processes occurring cannot be analysed selectively, since the possibility of side-reactions of primary fragments exists.

Alternative FPy–MS methods, in which thermal oligomers are collected in a reservoir before direct analysis (integrating method[18]), do not offer more advantages since the possibility of secondary reactions is enhanced and the probability of detection of high mass fragments is lessened by possible condensation in the reservoir, and hence 'loss' of such fragments.

Programmed temperature heating GPy–MS, is achieved using the direct insertion rod for solid samples (Fig. 5.1). The probe temperature is gradually increased on a linear programme and the evolving products are

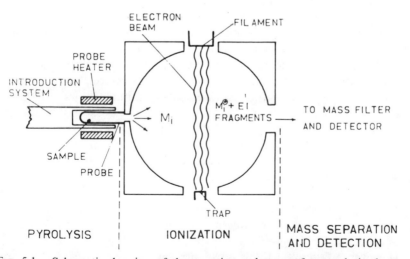

FIG. 5.1. Schematic drawing of the experimental set up for pyrolysis-electron impact mass spectrometry of polymers. M_i, pyrolytic products from the thermal decomposition of the polymer; $M_i^\oplus$, molecular ions of the pyrolytic products.

analysed by repetitive mass scans, almost simultaneously with their formation.[19] Using this method, the thermal energy is gradually transmitted to the polymer sample and it is possible to select the heating programme which permits the best time–temperature resolution of the thermal decomposition reactions.

The main advantage of the latter procedure is that, since pyrolysis is accomplished under high vacuum, thermal oligomers are volatilised and removed readily from the hot zone. This, together with the low probability of molecular collisions and fast detection, reduces to a great extent the occurrence of secondary reactions, so that almost exclusively primary fragments are detected.

Pyrolysis is obtained very close to the ion source and the resolving power of the mass filter is the only limiting factor in the detection of high mass fragments.

5.2.1.2. Ionisation

Among several alternative methods of ionisation available, those more currently employed are electron impact (EI), field ionisation (FI), field desorption (FD) and chemical ionisation (CI).

Electron impact is the oldest and still the most widely used method. In the EI ionisation mode, molecules are ionised by bombardment with an electron beam emitted by a heated tungsten wire.

An electron energy of 70 eV is commonly used. Since first ionisation energies of organic molecules are in the order of 9–12 eV, a considerable excess of energy is transmitted to the ionised molecules. This produces several fragment ions through unimolecular dissociation or rearrangement of the ionised molecules. The use of lower (15 eV) electron energy reduces considerably the formation of EI fragments, yielding simpler spectra.

In the case of FI, the molecules are strongly polarised in the inhomogeneous field around a platinum wire and drawn to the surface of this wire. The ionisation occurs either by tunnelling of an electron or by protonation of the species by absorbed proton donors (H_2O); after this, the positive ion is pushed out of the source and drawn into the accelerating stage.

In comparison with EI, FI is a 'softer' ionisation mode, and therefore molecular ions are mainly produced, and the formation of fragment ions is greatly reduced.

However, FI shows a lower sensitivity with respect to EI and a poor reproducibility due to the state of the activation of the anode.

In the FD technique, solid samples to be ionised are adsorbed on the surface of an activated emitter. The emitter is constituted with a $10 \mu m$ diameter tungsten wire which is treated at $1200\,°C$ with benzonitrile. This treatment produces a forest of microneedles with lengths up to $10 \mu m$ on the surface of the tungsten wire, thus enhancing the applied field to field ionisation levels.

The samples can be deposited on the emitter from solutions. After evaporation of the solvent the emitter is heated electrically until, at the same time, the material is desorbed and ionised. In the case of polymers concomitant thermal degradation reactions may be produced. Field desorption has been used in the investigation of bio-macromolecules.[20] Its application to synthetic polymers is yet to be exploited.

The CI technique consists in allowing a 'reaction gas', e.g. methane, to undergo an ion–molecule reaction in the MS ion source at a relatively high pressure. The ion formation process occurs with very little energy transfer and, consequently, simplified spectra are also produced here with respect to EI. A preliminary study has been recently reported[21] on the use of CI in the direct MS analysis of polymers, and it has been concluded that it can give useful information for polymer identification.

5.2.1.3. Mass Separation and Detection
Commercial spectrometers can be divided in to low and high resolution instruments. (High resolution is necessary in order to obtain the elemental composition of the ions.) A crucial aspect of mass separation is the ability to detect high molecular weight ions. Since commercial spectrometers are directed to analyse mostly low molecular weight compounds, their resolving power may exceed a hundred thousand in the high resolution instruments, but the mass range that they can explore is limited (frequently up to mass 1000 as an upper limit). Higher masses can usually be detected by decreasing the accelerating voltage. However, this also results in a loss of sensitivity.

On the other hand, the informative power of characteristic fragments evolving from a decomposing polymer is a function of the fragment mass. As a rule, low mass fragments carry a low amount of structural information about the original polymer structure, while fragments with high masses are generally the most informative. With the accumulation on MS data of polymers, it is becoming increasingly clear that terminal fragments volatising from polymers decomposing into the spectrometer may be of considerable size, so that they may remain undetected because of the limited mass range of many mass spectrometers. Fragments with masses of 2000–6000 are not uncommon in fluoropolymers,[13] while masses over ten thousand have been recently detected in polystyrenes.[22] Mass spectrometers having a mass range up to 4000–6000 are already commercially available, and their usefulness will become apparent in the near future.

Detection of high mass peaks is complicated by several factors. The first is the current technical limitations in the resolving power obtainable; the second is connected with the absence of suitable mass references above 2000 amu. One of the possible spin-offs from MS studies of polymers has been that they provide such reference standards.[22]

In addition to this, in high molecular weight compounds, the positive mass defect of hydrogen should be taken into account. In fact, in a high mass fragment, the very large accumulations of mass defect value may result in a shift of several mass units with respect to the nominal value.

The variable composition of the thermal fragments evolved in the Py–MS analysis of polymers makes it desirable to use the repetitive scanning technique, and thus the conventional multiplier detector appears to be preferable to the alternative photographic plate for mass detection.

5.2.1.4. Compound Identification in the Mixture of Thermal Degradation Products

Independently from the pyrolytic and ionisation methods employed, the main problem connected with the mass spectral analysis of polymers is the identification of products in the complex multicomponent mixture produced by thermal degradation. In fact, in the overall-end spectrum of a polymer, the molecular ions of the thermally formed oligomers will appear mixed with the fragment ions formed in the ionising step. Therefore, an interpretation of the mass spectral data requires a clear differentiation between the thermal degradation processes and those induced by ionisation.

This problem is minimised when soft ionisation methods, e.g. FI, FD or CI, are used, but becomes dramatic using EI at the usual electron energy of 70 eV. The use of a reduced electron energy (13–18 eV) produces a general decrease of the EI fragmentation, thus yielding less complicated spectra.

However, the reduction of the ionisation cross-section also reduces sensitivity; in addition, the information that may eventually be gained by the analysis of the EI induced fragmentation is lost. Consequently, it appears that the comparison of spectra taken at normal (70 eV) and reduced (13–18 eV) electron energies is the most useful approach in the elucidation of the EI mass spectra of polymers.

It has been shown[19] that an investigation of the EI fragmentation pathways of low molecular weight model compounds, reproducing characteristic structural entities of the polymer chain, may give useful information allowing us to differentiate between the EI induced and thermal degradation steps. Another possibility is to compare the mass spectrum of an authentic sample with that obtained directly from the polymer.

The use of high resolution instruments allows the elemental composition of ions to be determined, and this constitutes basic information for structure assignments to thermal oligomers. However, the usefulness of high-resolution MS analysis may be limited by the inability to separate and identify isomers, unless additional information is obtained from metastable processes or fragment, or double charged ions.

A promising MS technique for the identification of the chemical identity of the components is the mixture of pyrolysis products in the collisional activation (CA) mass spectrometry. If ions of high kinetic energies (>1 keV) collide with neutral target atoms, part of the translational energy is converted to internal energy, leading to subsequent collision-induced fragmentations. If the collision occurs in the second field-free region of a double-focusing instrument of reversed geometry, the collision-induced fragments may be energy analysed, then mass analysed conveniently by scanning the electric sector. The fragmentation pattern of such a CA spectrum on a given molecular ion qualitatively resembles that obtained by electron impact.[23]

The close similarity between the fragmentation patterns of EI and CA spectra allows the identification of the components of a mixture by matching the CA spectrum of the unknown compound with that of a known reference compound, or by interpretation of the fragmentation pattern. By selecting the molecular ions that characterise the various components of the mixture in the normal mass spectrum and successively recording the CA spectra of such peaks, it should be possible to identify these compounds without prior separation. However, in the application of this technique it appears necessary to produce the primary spectrum of the mixture under conditions that favour the occurrence of molecular ions. The method has been successfully employed for the identification of the components of a mixture produced by pyrolysis of biomaterials.[23]

5.2.2. Data Handling
The character and the quality of the data obtained in the direct MS analysis of polymers is to a great extent determined by the pyrolytic method employed.

Since flash pyrolysis (FPy) provides an instantaneous production of pyrolytic fragments, a single mass spectrum recorded at the optimal decomposition temperature will define the mass spectral pattern of the polymer. These spectra are, in general, highly reproducible and can be used as 'fingerprints' in polymer identification.[11] Flash pyrolysis has been used in combination with FDMS and low energy EIMS, in order to minimise the production of fragment ions, and to help in the recognition of the molecular ions in the complex mixture of thermal oligomers formed.

With gradual heating (GPy) the mass spectral pattern may be constantly changing as the temperature is raised. This phenomenon may have different origins. In one case, it occurs when the polymer system contains residual monomers, initiators, additives and oligomers formed in the polymerisation reaction. These components have, in general, different volatilities and may distill separately on raising the probe temperature. On the other hand, changes in the mass spectral pattern are observed for polymers which degrade thermally stepwise, since in this case different fragments are generated at different temperatures.

This represents a basic difference with respect to the mass spectral analysis of low molecular mass organic compounds, which generally volatilise undecomposed. Consequently, repetitive mass scans are necessary during the heating in order to detect all volatile products being evolved. The use of on-line computers is most valuable for the storage and handling of these data.

The mass spectral analysis of polymers by gradual heating may require a time period which may run into several minutes or even hours, and therefore a remarkably stable, reproducible and sensitive ionisation method must be used. These requirements are fulfilled by EI ionisation, which consequently has been largely used in combination with a programmed heating system. Combination of linear heating and EI in the mass spectral analysis of polymers, allows total ion current (TIC) and single ion current (SIC) curves to be obtained.

Total ion current curves are related to the total amount of volatile products evolved from a polymer upon heating. It has been pointed out[24] that, using a linear heating rate, TIC curves reproduce, qualitatively, the differential thermogravimetric curves, so that their maxima correspond to the maximum decomposition rate of the polymer in the high vacuum of the MS.

The correlation between TIC curves and differential TG is shown in Fig. 5.2, where thermogravimetric curves obtained for some polyvinyls are compared with the corresponding TIC curves.

Single ion current curves are obtained from the spectra recorded at

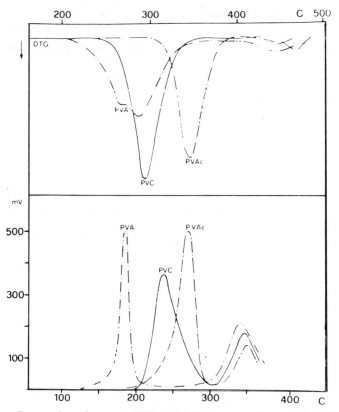

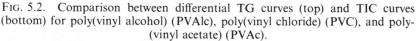

FIG. 5.2. Comparison between differential TG curves (top) and TIC curves (bottom) for poly(vinyl alcohol) (PVAlc), poly(vinyl chloride) (PVC), and poly-(vinyl acetate) (PVAc).

increasing temperatures, taking the absolute ion intensity of the more significant mass peaks associated with a given degradation product, and plotting this intensity versus temperature. These curves are also called 'fragmentograms'.

Provided that the peaks chosen do not suffer from contributions from other fragments these curves are related to the evolution of single thermal oligomers during the temperature raise.

The information value of the 'fragmentograms' is relevant, since pyrolytic products formed in the same thermal degradation reaction show parallel evolution profiles.

Therefore, an analysis of the SIC curves allows one to differentiate between the thermal reactions occurring in the polymer and to relate groups of thermal fragments to a specific process. Furthermore, these curves can be used to derive kinetic parameters (discussed in the next section).

5.2.3. Mass Spectrometric Thermal Analysis

As mentioned above, the linear programmed thermal degradation of polymers into the MS resembles differential thermogravimetric analysis (DTGA), and therefore the data obtained by MS can be elaborated in order to extract the usual thermoanalytical kinetic parameters (activation energy and pre-exponential factor).

Mass thermal analysis (MTA) presents several advantages with respect to the classical TGA isothermal and non-isothermal degradation methods.[25,26,27]

In fact, thermogravimetric data are not particularly informative in most cases of polymer degradation, when several degradation species are formed simultaneously or sequentially. In these cases one observes, at each temperature, the overall-end result of a superimposition of reactions. This is also true with MS if the total ion current (TIC) is used to monitor the polymer degradation. Instead, SIC curves permit each degradation process occurring in the thermal decomposition to be followed separately (Fig. 5.3).

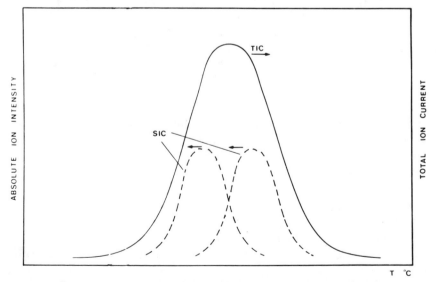

FIG. 5.3.   Time–temperature resolved pyrolysis-MS of polymers.

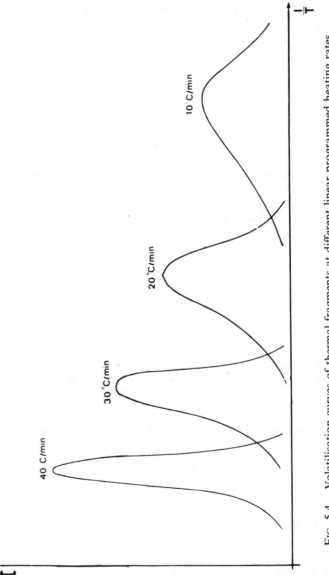

FIG. 5.4.   Volatilisation curves of thermal fragments at different linear programmed heating rates.

Therefore, it is possible to attach a molecular significance to the kinetic parameters eventually derived.

Using MTA, one is limited, among the thermal analytical methods available to calculate activation energies and pre-exponential factors, to a non-isothermal method which does not make use of initial weight loss data in the calculations. In fact, these quantities are hardly measurable using standard MS equipment.

Actually, it has been shown theoretically,[28,29] that it is possible to derive meaningful kinetic parameters from linear heating MS experiments. The method is based on the generation of a family of volatilisation curves obtained degrading a polymer sample at different linear programmed heating rates (Fig. 5.4).[28,29]

The high potential of MTA in elucidating the mechanisms of thermal decomposition of polymers has been recognised several times[25,26,28a,b,29] but it remains largely unexploited up to now.

Recent MTA studies[30] have shown that the activation energy values obtained by the method of different linear programmed heating rates[28,29] for PMMA and α-Me-PS (decomposing by the irreversible unzipping mechanism), are in excellent agreement with the values obtained by running parallel TGA experiments at ambient pressure.

5.3.   APPLICATIONS

5.3.1. Structure Identification and Thermal Degradation Mechanisms

The mass spectral characterisation of polymers and copolymers by MS leads to their structure identification and yields also pertinent information on mechanisms of thermal degradation.

In the following, the applications of MS to polymers are discussed with the purpose of illustrating significant examples, rather than giving a comprehensive account.

As mentioned above, the use of EI to reveal thermal fragments has been preponderant over other ionisation techniques. However, we shall discuss the results independently from the particular kind of ionisation in the MS investigation.

5.3.1.1. Polyesters

A large number of polyesters and polylactones has been investigated by GPy–EIMS.[31–32] The polymers derived from dicarboxylic acids and glycols or hydroquinone can be divided into aliphatic polyesters, and aliphatic–aromatic polyesters. Aliphatic polyesters include polymers from succinic or adipic acid and ethyleneglycol, 1,3-propanediol, 1,4-butane-diol, 2,2'-dimethyl-1,3-propanediol and 2,2'-oxydiethanol.

These polymers decompose in the MS between 250 and 280 °C and the occurrence of two different thermal degradation mechanisms has been inferred from the MS data.[33]

The 70 eV spectrum of poly(ethylene adipate), recorded at a probe temperature of 280 °C, is shown in Fig. 5.5, together with that of poly(d_4-ethylene adipate) obtained under the same conditions.

The spectrum contains a series of intense peaks at m/e 173, 345 and 517, which might be explained postulating a thermal cleavage of the ester bond with concomitant hydrogen transfer and subsequent EI fragmentation into carboxonium ions (eqn. (5.1)).

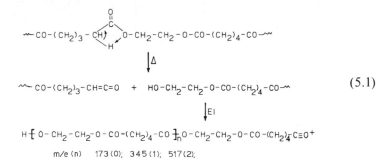

$$(5.1)$$

From the same thermal fragments an alternative EI fragmentation produces the ions at m/e 217 and 389 (eqn. (5.2)).

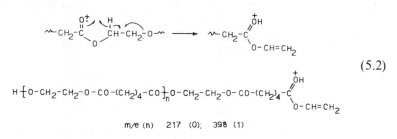

$$(5.2)$$

The mass spectrum of the deuterated analogous poly(d_4-ethylene adipate) confirms that the hydrogen transfer occurs from the aliphatic dicarboxylic acid (Fig. 5.5).

Instead, a thermal degradation reaction involving an intramolecular β-hydrogen transfer from the aliphatic chain of the glycol (*cis* elimination)

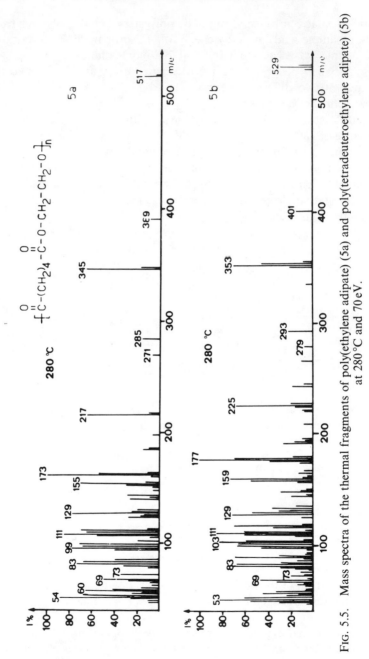

FIG. 5.5.   Mass spectra of the thermal fragments of poly(ethylene adipate) (5a) and poly(tetradeuteroethylene adipate) (5b) at 280°C and 70eV.

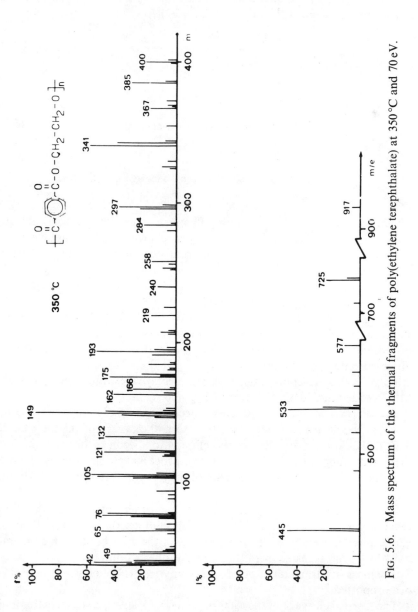

FIG. 5.6.   Mass spectrum of the thermal fragments of poly(ethylene terephthalate) at 350°C and 70 eV.

and subsequent EI fragmentation to carboxonium ions explains the less intense peaks at m/e 129 and 155 (eqn. (5.3)).

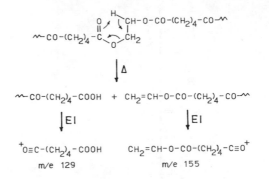

(5.3)

The mass spectra of the remaining polymers of this family reveal that the transfer of hydrogen from the dicarboxylic acid with concomitant cleavage of the ester bond is the predominant thermal degradation process, whereas the alternative *cis* elimination produces less intense ions.[33]

The aromatic–aliphatic polyesters studied are essentially derived from phthalic acid and various aliphatic diols.[34]

Poly(ethylene terephthalate) and poly(tetramethylene terephthalate) show enhanced thermal stability with respect to the aliphatic polyesters. For these polymers, the evolution of thermal degradation products is observed only at probe temperatures higher than 340 °C.

Poly(ethylene terephthalate) is a representative example and its 70 eV MS recorded at 350 °C is shown in Fig. 5.6.

The thermal degradation processes occurring in these polymers appear to be more selective. In fact, since the thermal cleavage of the ester bond (eqn. (5.1)) cannot take place, the *cis* elimination becomes the predominant thermal degradation mechanism, and this process explains the most intense series of peaks found in the spectrum of poly(ethylene terephthalate) (eqn. (5.4)).

However, less intense peaks suggest that other thermal degradation reactions are also occurring. These include elimination of carbon dioxide, elimination of acetaldehyde, and cleavage of the ester bond with concomitant hydrogen transfer from the adjacent glycol unit.

The mass spectral data of poly(oxysuccinyloxy-1,4-phenylene) have also been reported.[35]

The thermal cleavage of the ester bond (eqn. (5.1)) is the selective thermal degradation reaction in this polymer.

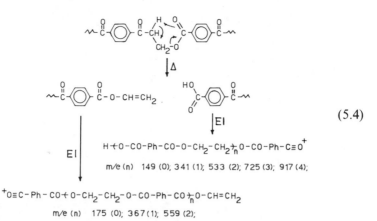

(5.4)

$$H + O-CO-Ph-CO-O-CH_2-CH_2 \overset{+}{\underset{n}{}} O-CO-Ph-C \equiv O^+$$

m/e (n)   149 (0); 341 (1); 533 (2); 725 (3); 917 (4);

$$^+O \equiv C-Ph-CO + O-CH_2-CH_2-O-CO-Ph-CO \overset{+}{\underset{n}{}} O-CH=CH_2$$

m/e (n)   175 (0); 367 (1); 559 (2);

In addition, it is found to have a peak at m/e 576 which is attributed to the molecular ion of a cyclic trimer (eqn. (5.5)).

$$\begin{bmatrix} + CO-CH_2-CH_2-CO-O-Ph-O \overset{}{\underset{2}{}} \\ -O-Ph-O-CO-CH_2-CH_2-CO \end{bmatrix}$$

(5.5)

Polylactones are thermally less stable than aliphatic polyesters from dicarboxylic acid and diols and their EIMS were produced at a probe temperature of 210–250 °C.[36–39] In some instances it was observed that the thermal degradation reactions occurring in the high vacuum of the mass spectrometer are different from those produced on heating at atmospheric, or reduced, pressure. One example is given by poly(1,1-oxycarbonyl-ethylene) and poly(oxycarbonylmethylene).[36]

Although poly-α-lactones are known to depolymerise on heating into six-membered cyclic dimers, the mass spectra of these polymers show peaks at mass values by far exceeding the molecular weight of the cyclic dimers.

The mass spectrum of poly(1,1-oxycarbonylethylene) is shown in Fig. 5.7.

A metastable transition shows that the base peak (m/e 56) is an EI fragment formed from the molecular ion of the cyclic dimer (m/e 144) by loss of carbon dioxide and acetaldehyde (eqn. (5.6)).

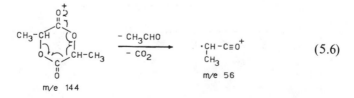

(5.6)

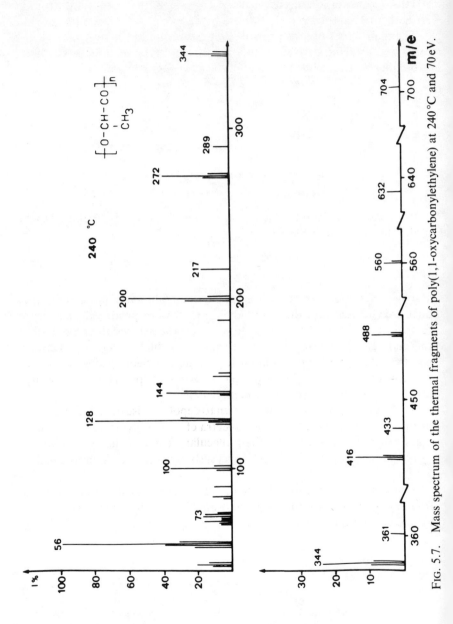

FIG. 5.7. Mass spectrum of the thermal fragments of poly(1,1-oxycarbonylethylene) at 240°C and 70eV.

This fragmentation mechanism was confirmed by the mass spectrum of an authentic sample.

The main series of ions present in the spectrum of the polymers can be explained as being derived from a mixture of cyclic oligomers which are subsequently fragmented by the mechanism shown in eqn. (5.7).

Polymer

$$m/e(n) \ 56 \ (0); \ 128 \ (1); \ 200 \ (2); \ 272 \ (3); \ 344 \ (4);$$
$$416 \ (5); \ 488 \ (6); \ 560 \ (7); \ 632(8); \ 704 \ (9);$$

(5.7)

A study of appropriate metastable transitions showed that these fragment ions further decompose by an EI fragmentation reaction which resembles a cationic depolymerisation process (eqn. (5.8)).

$$\cdot CH-CO-O-CH-CO-O-CH-CO \!\!\! \mid \!\! O-CH-C\!\equiv\!O^+$$
$$\quad | \qquad\qquad | \qquad\qquad | \qquad\qquad |$$
$$\quad CH_3 \qquad\quad CH_3 \qquad\quad CH_3 \qquad\quad CH_3 \quad m/e \ 272$$

$$\cdot CH-CO-O-CH-CO \!\!\! \mid \!\! O-CH-C\!\equiv\!O^+$$
$$\quad | \qquad\qquad | \qquad\qquad |$$
$$\quad CH_3 \qquad\quad CH_3 \qquad\quad CH_3 \quad m/e \ 200$$

(5.8)

$$\cdot CH-CO \!\!\! \mid \!\! O-CH-C\!\equiv\!O^+$$
$$\quad | \qquad\qquad |$$
$$\quad CH_3 \qquad\quad CH_3 \quad m/e \ 128$$

$$\cdot CH-C\!\equiv\!O^+$$
$$\quad |$$
$$\quad CH_3 \quad m/e \ 56$$

The thermal cleavage of the ester bond (eqn. (5.1)) produces less intense ions:

$$H \!\!\! \mid \!\! O-CH-CO \!\!\! \mid_n O-CH-C\!\equiv\!O^+$$
$$\qquad\quad | \qquad\qquad |$$
$$\qquad\quad CH_3 \qquad\quad CH_3$$

(5.9)

$$m/e \ (n) \ 73 \ (0); \ 145 \ (1); \ 217 \ (2); \ 289 \ (3); \ 361 \ (4); \ 433 \ (5);$$

The mass spectrum of poly(oxycarbonylmethylene) was interpreted by analogous considerations.[36]

Poly(oxycarbonylethylene) is thermally degraded mainly by *cis* elimination[37] (eqn. (5.3)).

The mass spectral investigation of pure pivalolactone provided the basis for the interpretation of the MS spectrum of poly(pivalolactone).[37,38]

In the mass spectrum of the monomer the nuclear ion (m/e 100) is absent and two main fragmentation pathways are found (eqn. (5.10)).

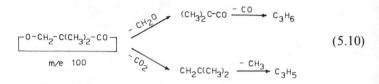

$$(5.10)$$

The mass spectrum of the polymer shows a series of fragment ions that can be considered as being derived from cyclic oligomers by the EI fragmentation reactions in eqn. (5.10). Although the molecular ions of the cyclic oligomers were totally absent, the formation of these compounds was confirmed by a separate pyrolysis of the polymer.[37]

Poly(oxycarbonyltetramethylene) gives on heating quantitative yields of the six-membered ring δ valerolactone, and the molecular ion of this thermal oligomer (m/e 100) is the highest mass peak found in the spectrum.[37]

In poly(oxycarbonylpentamethylene) the cleavage of the ester bond (eqn. (5.1)) was the selective thermal degradation mechanism found under the experimental conditions occurring in the MS (high vacuum).[39]

5.3.1.2. Polyamides and Polyimides

Several commercial aliphatic polyamides derived from dicarboxylic acid and diamines (nylon-6,6,10, -4,10, -11,6, -12,12) have been investigated by GPy–EIMS.[40]

All these polyamides contained cyclic oligomers formed in the condensation reaction, which evaporated undecomposed in the MS and were detected at probe temperatures below 200 °C (see Section 5.3.4.1).

Evolution of thermal fragments was observed at probe temperatures above 350 °C.

The thermal degradation breakdown of these polymers parallels that of the corresponding polyesters. In fact, the cleavage of the amide bond and the *cis* elimination were the thermal degradation processes ascertained (eqns. (5.11) and (5.12)).

Nylon-6,6 has also been investigated by FPy at 600 °C and FI.[41]

The FI spectrum obtained under these conditions showed peaks

$$\text{\textasciitilde}CH_2-CO-NH-(CH_2)_n-NH-CO-(CH_2)_m-CONH_2 \;+\; CH_2=CH-(CH_2)_{m-2}NH-CO\text{\textasciitilde}$$

$$\Big\uparrow\Delta \qquad\qquad (5.11)$$

$$(5.12)$$

$$\Big\downarrow\Delta$$

$$H_2N-(CH_2)_n-NH-CO-(CH_2)_m-CO-NH-(CH_2)_n-NH-CO\text{\textasciitilde}$$

attributable to amide fragments, identical to those found in the GPy–EI experiment, but suggested also the formation of water, ammonia, cyclohexane or hexene, hexamethylenediamine, hydrogen cyanide and nitriles, as secondary pyrolytic products formed by further thermal degradation of primary thermal fragments. In the same paper[41] the mass spectra (obtained by FPy at temperatures of 600–700 °C and FI), of aromatic polyamides (derived from terephthalic acid) and different diamines (p-phenylenediamine, 4,4'-bisphenylenediamine, 4,4'-oxydianiline) are discussed. Amines, amides, isocyanates, nitriles are characteristic thermal fragments found under those conditions.

From these findings it was concluded that, in agreement with the results reported above (eqns. (5.11) and (5.12)), the CH$_2$—NH linkage is the weakest bond within the chain.

Other investigations have been concerned with polyamides of several cycloaliphatic dicarboxylic acids and piperazine,[42] and truxillic and truxinic acids and piperazine[43,44] (see Section 5.3.2).

The mass spectra of some poly-β-propiolactams and nylon-4, -6, -7, and -12, obtained by GPy–EIMS, have been reported.[19,40,45-47]

The mass spectra of poly-β-propiolactams recorded at probe temperatures of 340–420 °C showed that these compounds decompose selectively by a thermal cis elimination, as the analogous polylactones[19,45,46] (eqn. (5.13)).

$$\text{\textasciitilde}HN-CR_2-CHR-C \xrightarrow{\;\;\Delta\;\;} CR_2=CR-CO-(NH-CR_2-CHR-CO)_n-NH_2, \qquad (5.13)$$

$$R=H,\; CH_3,\; C_2H_5.$$

The other polylactams studied decompose essentially by formation of cyclic oligomers.

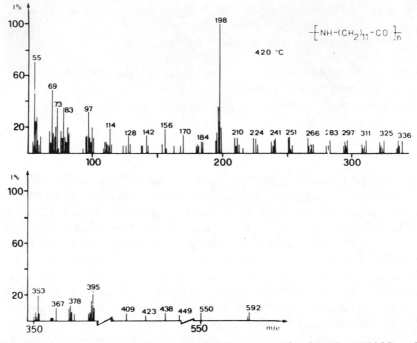

FIG. 5.8.　Mass spectrum of the thermal fragments of nylon-12 at 420 °C and 70 eV.

Nylon-6, -7, and -12 give, at probe temperatures of 350–430 °C, a mixture of cyclic monomer, dimer, trimer and tetramers.[40,47]

The MS of nylon-12 at 420 °C is shown in Fig. 5.8, as an example. It shows the molecular ions of the cyclic monomer, dimer and trimer at m/e 197, 394 and 591 and additional EI fragments at m/e 198, 395, 592 formed by a fragmentation reaction typical for oligomeric cyclic lactams (eqn. (5.14)).

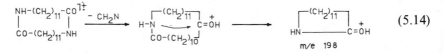

Nylon-6 was also investigated by FPy (600 °C) and FIMS.[41] Under these conditions the cyclic monomer was almost exclusively the thermal degradation product.

The thermal degradation breakdown found for nylon-6 represents a remarkable difference with respect to its oxygen analogue, polycaprolactone, which degraded the MS to yield open chain fragments.

Nylon-4 is the less thermally stable polymer among the polylactams studied. Its MS at 300 °C allows us to ascertain the predominant formation of cyclic monomer accompanied by minor amounts of dimer.[40] The molecular ions of these products were not found in the mass spectrum which showed instead their EI fragments formed by the fragmentation reaction in eqn. (5.14).

An investigation by FPy–FIMS of thermally stable polyimides based on pyromellitic and trimellitic acid, has been reported.[41]

5.3.1.3. Polyurethanes, Polyureas and Polycarbonates

GPy–EIMS studies of polyurethanes have been carried out recently.[24,48,49] A heating rate of 10 °C min^{-1} was used; N-monosubstituted polyurethanes decompose between 300 and 350 °C at atmospheric pressure, but in the high vacuum of the MS the decomposition temperature is lowered to about 200 °C. These polymers undergo, quantitatively, a depolycondensation process, as depicted in eqn. (5.15).

$$\text{⌁ Ar}-\underset{\underset{H}{|}}{N}-\underset{\underset{O}{\|}}{C}-O-R\text{⌁} \quad \xrightarrow{\Delta} \quad \text{Ar}\!\!\begin{array}{l} \diagup N=C=O \\ \diagdown N=C=O \end{array} \quad + \quad R\!\!\begin{array}{l} \diagup OH \\ \diagdown OH \end{array} \qquad (5.15)$$

In Fig. 5.9 is shown the EI mass spectrum of the polyurethane from methylenediphenyldiisocyanate (MDI) and 1,4-butanediol, which appears to be essentially that of a mixture of MDI and butanediol.

Instead, the thermal decomposition of N-disubstituted polyurethanes

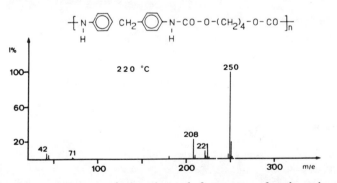

FIG. 5.9. Mass spectrum of the thermal fragments of polyurethane from methylenediphenyldiisocyanate (MDI) and 1,4-butanediol at 220 °C and 18 eV.

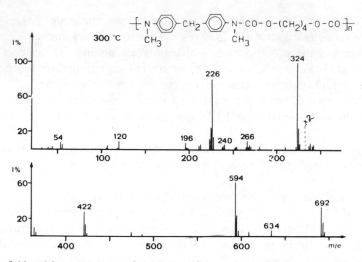

FIG. 5.10.   Mass spectrum of the thermal fragments of polyurethane from 4,4′-diaminediphenylmethane and 1,4-butanediol-bischloroformate at 300°C and 18 eV.

TABLE 5.1
Fragments from the Pyrolysis of Polyurethane from 4,4′-Diaminodiphenylmethane and 1,4-Butanediol-bischloroformate at 300°C

Fragment[⊕]	m/e
$\overset{CH_3}{\underset{\vert}{}}$ $\overset{CH_3}{\underset{\vert}{}}$ $\overset{CH_3}{\underset{\vert}{}}$ $\overset{CH_3}{\underset{\vert}{}}$ H—NPhCH₂PhN—CO₂—(CH₂)₄—CO₂—NPhCH₂PhN—CO₂—CH₂—CH₂—CH=CH₂	692
$\overset{CH_3}{\underset{\vert}{}}$ $\overset{CH_3}{\underset{\vert}{}}$ $\overset{CH_3}{\underset{\vert}{}}$ $\overset{CH_3}{\underset{\vert}{}}$ H—NPhCH₂PhN—CO₂—(CH₂)₄—CO₂—NPhCH₂PhN—H	594
$\overset{CH_3}{\underset{\vert}{}}$ $\overset{CH_3}{\underset{\vert}{}}$ CH₂=CH—CH₂—CH₂—CO₂—NPhCH₂PhN—CO₂—CH₂—CH₂—CH=CH₂	422
$\overset{CH_3}{\underset{\vert}{}}$ $\overset{CH_3}{\underset{\vert}{}}$ H—NPhCH₂PhN—CO₂—CH₂—CH₂—CH=CH₂	324
$\overset{CH_3}{\underset{\vert}{}}$ $\overset{CH_3}{\underset{\vert}{}}$ H—NPhCH₂PhN—H	226
CH₂=CH—CH=CH₂	54

occurs selectively through eqn. (5.16) (*cis* elimination) as demonstrated by the detection of thermal fragments containing secondary amine and olefinic end-groups[24] (Fig. 5.10 and Table 5.1).

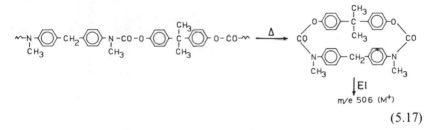

$$\tag{5.16}$$

Totally aromatic *N*-disubstituted polyurethanes are considerably more stable and decompose via the formation of a cyclic compound,[49] as shown in eqn. (5.17) and in Fig. 5.11.

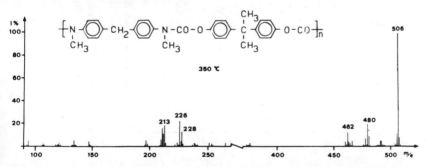

$$\tag{5.17}$$

The above mentioned results show that there are several primary fragmentation mechanisms which occur in the thermal decomposition of polyurethanes.

In the *N*-monosubstituted polyurethanes, the depolymerisation process involves a low energy hydrogen transfer from the nitrogen atom to yield di-isocyanate and di-alcohol primary fragments, and therefore it occurs selectively at relatively low temperatures.

FIG. 5.11. Mass spectrum of the thermal fragments of polyurethane from *N,N'*-dimethyl-4,4'-diaminodiphenylmethane and 4,4'-isopropylidenediphenol-bischloroformate at 350 °C and 16 eV.

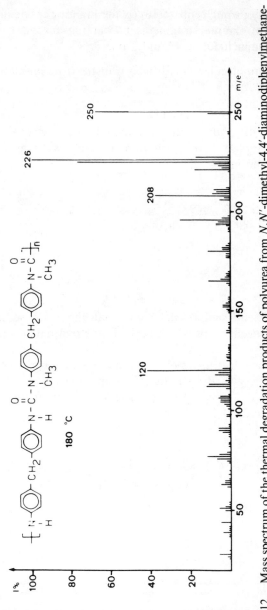
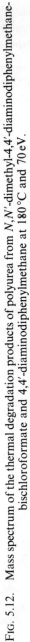

Fig. 5.12.   Mass spectrum of the thermal degradation products of polyurea from N,N'-dimethyl-4,4'-diaminodiphenylmethane-bischloroformate and 4,4'-diaminodiphenylmethane at 180 °C and 70 eV.

In the N-disubstituted polyurethanes, the low energy hydrogen transfer process seen above cannot occur, so that some higher energy processes may take place; these include:

(a)   the intramolecular β-hydrogen transfer from an aliphatic carbon atom (*cis* elimination) to form amine, olefine and carbon dioxide;
(b)   the formation of a cyclic compound by an intramolecular exchange reaction.

The cyclisation reaction is, in this case, a relatively high energy process, which occurs selectively because the polymer structure is such that hydrogen transfer reactions along the chain are not possible.

Polyureas, which are structurally very similar to polyurethanes, have also been investigated by GPy–EIMS.[50] In Fig. 5.12 is shown the EI spectrum of a N,N'-monosubstituted polyurea, which appears to be essentially that of a mixture of di-isocyanate and an amine. This demonstrates that these polymers undergo, quantitatively, a depolymerisation process as depicted in eqn. (5.18).

$$(5.18)$$

Polycarbonates are the last in this family of polymers. Many studies have been dedicated to the thermal decomposition of totally aromatic polycarbonates since they have relevant practical and industrial interest.[51] GPy–EIMS of poly(4,4'-isopropylidenediphenylcarbonate) has been performed[52] and the MS of the volatile decomposition products is reported in Fig. 5.13.

These data have been interpreted as showing the formation of a cyclic trimer (m/e 762) by an intramolecular exchange reaction.

5.3.1.4. *Polyethers and Polythioethers*
Polyoxymethylene is known to unzip to give the monomer when heated. The mass spectrum at 310 °C reveals, however, the presence of trioxane and higher cyclic oligomers.[53]

The mass spectral data relative to the pyrolysis products of polypropyleneglycol have been described[11] and the degradation has been explained by the cleavage of the C—O bond, formation of hydroxyl endgroup and a partial loss of water from these fragments.

In a study concerning some aromatic polyethers and polythioethers, the

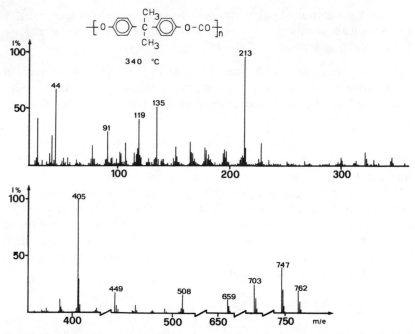

FIG. 5.13.   Mass spectrum of the thermal degradation products of poly(4,4'-isopropylidenediphenylcarbonate) at 340 °C and 70 eV.

polymers were pyrolysed by raising the probe temperature at 40 °C min^{-1} and a low electron energy (15 eV) was used in order to reduce the EI fragmentation.[54,55]

Statistical cleavage of the polymer chain and stabilisation by hydrogen transfer was found in poly(oxy-1,4-phenylene) (Fig. 5.14).

Fragments having one phenyl and one hydroxyl end-group, in agreement with their higher probability of formation, constitute the main series of peaks in the spectrum. They are surrounded by less intense peaks spaced by ± 16 amu, corresponding to oligomers with two phenyl or two hydroxyl end-groups.

Poly(oxy-2,6-dimethyl-1,4-phenylene) has been studied by GPy–EIMS[56] and poly(oxy-1,3-phenylene), poly(oxy-2,6-dimethyl-1,4-phenylene), poly-(oxy-2-methyl-6-phenyl-1,4-phenylene),   poly(oxy-2,6-diphenyl-1,4-phen-ylene), and poly(oxy-2,5-dimethoxy-1,4-phenylene) were investigated by FPy–FDMS.[18]

The main feature showed by the mass spectra is that the thermolysis of these polymers proceeds by statistical chain splitting.

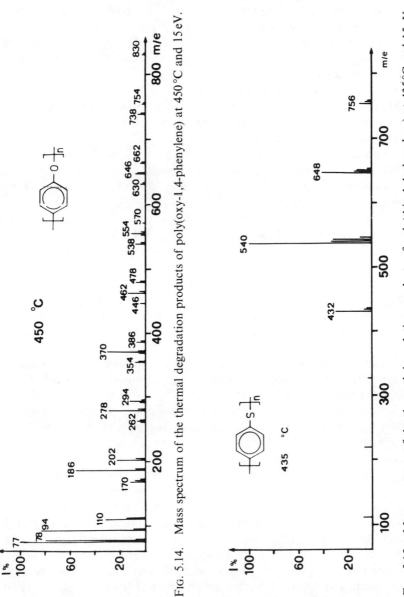

FIG. 5.14.   Mass spectrum of the thermal degradation products of poly(oxy-1,4-phenylene) at 450°C and 15 eV.

FIG. 5.15.   Mass spectrum of the thermal degradation products of poly(thio-1,4-phenylene) at 435°C and 15 eV.

In contrast with these results, poly(thio-1,4-phenylene) showed a very different mass spectral pattern (Fig. 5.15).[55,57] Here, only EI fragments corresponding to the molecular ions of oligomers, of general formula $+C_6H_5-S+_n$ with $n = 4$, 5, 6, 7, are found. Considering the high intensities of these fragments, it seems appropriate to attribute these peaks to cyclic oligomers.

Poly(dithio-1,4-phenylene) shows a continuously changing mass spectral pattern as the temperature is increased (Fig. 5.16).[55] The occurrence of different pyrolytic reactions was followed by means of mass 'fragmentograms' of selected ions (Fig. 5.17).

Oligomers corresponding to a cyclic dimer (m/e 280) and a tetramer (m/e 420) are evolved at low temperatures (260 °C, Fig. 5.17), whereas the expulsion of sulphur is recognised by the presence of a peak at m/e 64 (S_2), showing a maximum at about 320 °C.

Thermal fragments with deficiency in sulphur (with respect to the original polymer) are found above 300 °C (m/e 464, 540, 572, 648). The high intensities of these peaks and their mass values suggest that they correspond to cyclic oligomers.

Although expulsion of the disulphide bridges with formation of polyphenylene blocks might be postulated, the formation of cyclic products is consistent with expulsion of single sulphur atoms. The mass fragmentograms (Fig. 5.17) give a clear, time–temperature resolved evolution profile of sulphur and cyclic oligomers.

It can be observed that cyclic oligomers produced at intermediate temperature still contain disulphide bridges (m/e 464, 472), whereas at higher temperatures only sulphur bridges are present (m/e 540, 648).

Similar results have been obtained in other polymers containing disulphide bridges.[58,59]

5.3.1.5. Polyhydrocarbons

Pyrolysis-mass spectra of polyethylene, polypropylene, polyisobutene and polybutadiene have been reported.[11,60,61] They show the characteristic pattern of low molecular weight hydrocarbons. In contrast to polyethylene and polypropylene, whose thermal degradation proceeds statistically,[11,60] polyisobutene is found to undergo a distinct depolymerisation reaction where tertiary C—C bonds are preferably cleaved.[11,61]

Polystyrene[11,61,62] and poly-α-methylstyrene[63] have also been investigated. The former polymer decomposes followed by some repolymerisation, yielding mainly styrene monomer, minor amounts of dimer and trimer and also traces of tetramer and pentamer.[11,61,62]

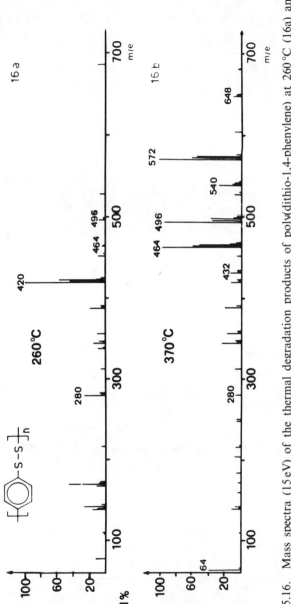

FIG. 5.16. Mass spectra (15 eV) of the thermal degradation products of poly(dithio-1,4-phenylene) at 260°C (16a) and 370°C (16b).

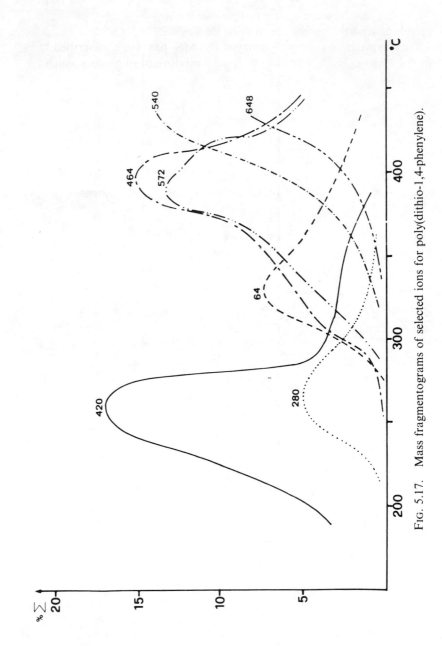

FIG. 5.17.   Mass fragmentograms of selected ions for poly(dithio-1,4-phenylene).

Poly-α-methylstyrene decomposes by the unzipping reaction, so that only monomer fragments are monitored by MS.[63]

The investigation of polybenzyls by MS has been described.[64,65] Polybenzyls yield, in the pyrolytic step, a mixture of oligomers which are strongly fragmented by EI.

5.3.1.6. Polyvinyls

Mass spectral data obtained by GPy–EIMS of some polyvinyls have been reported.[11,66] It is interesting to note that the mass spectrum of PVC taken at 230 °C differs considerably from that taken at 340 °C (Fig. 5.18). This behaviour is typical for vinyl polymers with a pendant electronegative X group. In fact, their thermal degradation involves the elimination of HX, as the first decomposition step, leading to a macromolecular polyene structure. At higher temperatures, the latter decompose and rearrange to yield sizeable amounts of aromatic hydrocarbons. Another example is given by poly(vinyl acetate)[66] (Fig. 5.19).

In the spectrum at 280 °C are present peaks at m/e 60, 45, 43, due to the

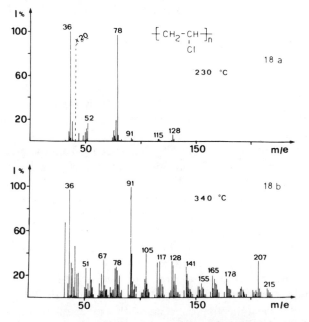

FIG. 5.18. Mass spectra (70 eV) of the thermal degradation products of poly(vinyl chloride) at 230 °C (18a) and 340 °C (18b).

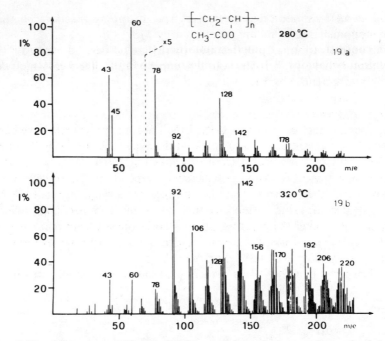

FIG. 5.19.   Mass spectra (18 eV) of the thermal degradation products of poly(vinyl acetate) at 280 °C (19a) and 320 °C (19b).

acetic acid evolving at this temperature from the polymer. Other intense peaks appear at m/e 78 and 128, corresponding to benzene and naphthalene, respectively.

The spectrum recorded at 320 °C differs sharply from the preceding one. Peaks due to the acetic acid are almost absent, while the evolution of aromatic compounds dominates.

The volatilisation rate profiles (total ion current against temperature) relative to three vinyl polymers have been shown in Fig. 5.2.

Each curve shows two peaks; the first peak occurs at different temperatures, depending on the dissociation energies relative to each R—X bond, while the second peak is centred at about 350 °C in all cases, and corresponds to the evolution of aromatic hydrocarbons.

The detailed information obtained in these pyrolytic MS studies have helped to clarify some fine aspects of the mechanism of thermal degradation of PVC and related polymers.[66,67]

Poly(vinylidene chloride) has also been studied.[11,68] The formation of halogenated aromatic compounds has been ascertained among the pyrolysis products.

In contrast to this, poly(tetrafluoroethylene) shows a uniform degradation behaviour.[11] It starts to decompose in the spectrometer above 500 °C and to yield, nearly quantitatively, the monomer. The dimer occurs only in low concentration.

Mass spectra of polyacrylonitrile have been reported[63] and the thermal behaviour of this type of polymer has been characterised in comparison to that of its copolymers.[69,70]

Vinyl polymers with ester and amide side groups exhibit different degradation reactions. Whereas poly(methyl methacrylate) depolymerises nearly quantitatively via a 'zip' mechanism into the monomer,[53] poly-(methyl methacrylamides) are degraded by a two step reaction.[71,72] At temperatures above 230 °C, a cyclisation with formation of imide structures can be observed; these at higher temperatures are cleaved off, through the unzipping reaction, to yield divinyl units (eqn. (5.19)).

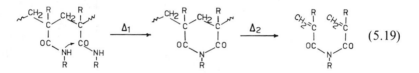

$$(5.19)$$

A very clear example of this behaviour is given by the poly(p-benzenesulphamidomethacrylamide).[71,72]

5.3.1.7. Biopolymers

At present, only a few examples of Py–MS studies on biological polymers have been reported. DNA was investigated by EI[73] and FDMS.[74]

In the latter case a suspension of herring DNA in acetone was transferred on the emitter by the dipping technique. The tungsten wire emitter was electrically heated and a submicropyrolysis on the surface of the emitter was achieved. The pyrolytic products were at the same time ionised and desorbed (Py–FDMS).

The high resolution mass spectral data were subsequently used for accurate mass measurements. This was greatly facilitated by the appearance of inorganic phosphate cluster ions throughout the Py–FDMS, which were used as internal mass references. The quasi-molecular ions (M + 1) of the five essential bases, adenine, guanine, cytosine, thymine and methylcytosine, of herring DNA were found in the spectrum. Further peaks at higher masses were identified as being due to fragments containing bases

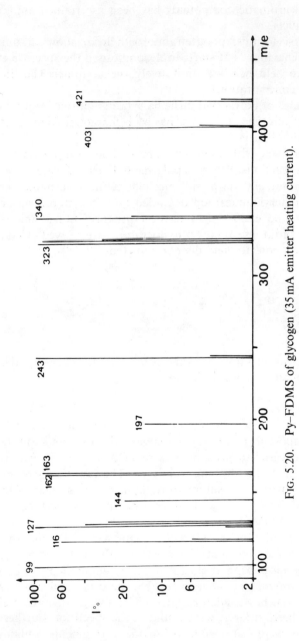

FIG. 5.20. Py–FDMS of glycogen (35 mA emitter heating current).

associated with phosphoric acid or chemically bound to dehydrated deoxyribose units.

Interestingly, quasi-molecular $(M + 2)$, doubly charged ions corresponding to dinucleotide fragments were found in the FD spectrum, whereas they were absent in the EIMS.

This result could be of great importance, in view of the possibility of obtaining base sequences information by Py–FDMS.

DNA and RNA have also been studied by Curie-point pyrolysis and low-voltage EI or high resolution FIMS.[75]

The results showed that the spectra from both techniques are in good agreement. Since most mass peaks correspond to pyrolytic fragments originating from the sugar moiety of the nucleic acid, the mass spectra can be used to differentiate DNA and RNA. The chemical identity of some peaks was confirmed by collisional activated (CA) spectra[23] and an explanation of the pyrolytic processes was given.

Studies of polysaccharides and polypeptides by Py–FDMS have been undertaken in order to clarify some of the reaction mechanism in Py–FDMS as well as to explore the potential of the method for structural analysis of biopolymers.[76,77]

The Py–FDMS of glycogen is reported in Fig. 5.20 as an example. It shows molecular and quasi-molecular ions derived from mono- $(m/e$ 162, 163) and di-saccharide units $(m/e$ 324, 325) (see eqn. (5.20)).

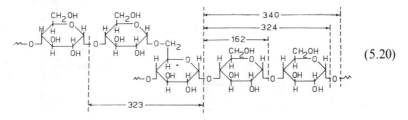

$$(5.20)$$

Other signals are due to compounds generated by pyrolytic rupture, water elimination and subsequent protonation of the dehydrated molecules.

5.3.1.8. Inorganic Polymers

Relatively little work has been carried out on the pyrolysis of inorganic polymers into the MS. In this section only two examples will be discussed; polyphosphazenes[78] and polysiloxanes.[11]

The mass spectral data show that the pyrolytic breakdown of poly-(dinaphtoxyphosphazene) (Fig. 5.21) leads to the formation of cyclic

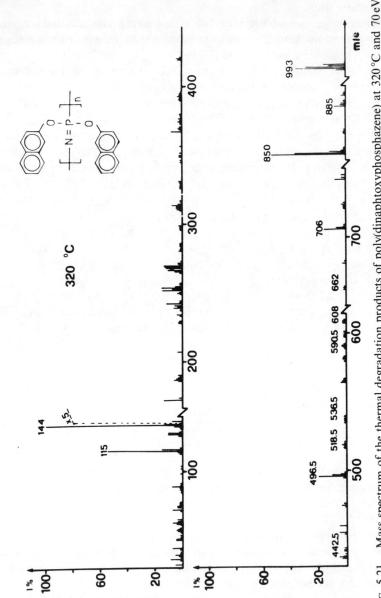

FIG. 5.21. Mass spectrum of the thermal degradation products of poly(dinaphtoxyphosphazene) at 320°C and 70eV.

oligomers. The cyclisation reaction, depicted in eqn. (5.21), is favoured
here because hydrogen transfer processes along the polymer main chain are
not possible.

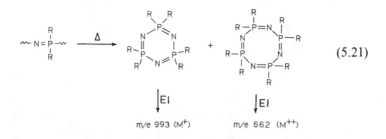

$$m/e\ 993\ (M^+) \qquad m/e\ 662\ (M^{++})$$

Contrary to the previous case, the thermal degradation of poly-
(dianilinephosphazene) begins at very low temperature and the only
volatile product detected is aniline.

The thermal degradation occurs in two stages, one around 180 °C and the
second at 290 °C. According to eqn. (5.22), aniline is evolved at first through
an intermolecular hydrogen transfer process involving two pendant amino
groups. This produces cross-linked phosphazene structures containing
tertiary amino groups, which can be removed by further heating, when the
charring of the cross-linked phosphazene structures is beginning.

Finally, the thermal decomposition of poly(dipiperidine phosphazene)
leads to the complete destruction of the polymer backbone, with the
formation of ammonia and elemental phosphorus.

The mass spectral data relative to this polymer are shown in Fig. 5.22.
The products evolving at 240 and 340 °C are markedly different. The MS at
240 °C indicates that piperidine (m/e 85, M+) and tetrahydropyridine
(m/e 83, M+) are the pyrolytic products that are evolved at this
temperature.

In the spectrum at 340 °C the base peak at m/e 124 is attributed to
elemental phosphorus P_4, the peaks at m/e 79 (M+) and m/e 52, 51 and 50
reveal the presence of pyridine, and the presence of ammonia and elemental
nitrogen is strongly suggested by the intense peaks at m/e 17 and m/e 28.

The mass fragmentograms for these species are reported in Fig. 5.23. It
can be noted that pyridine, phosphorus and ammonia increase when

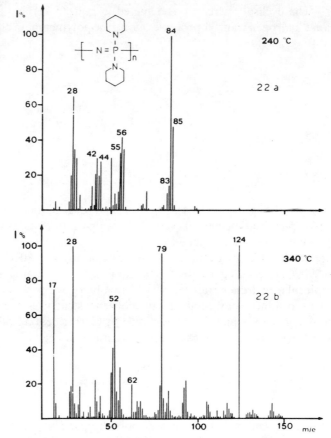

FIG. 5.22.   Mass spectra (70 eV) of the thermal degradation products of poly-(dipiperidinephosphazene) at 240 °C (22a) and 340 °C (22b).

piperidine and tetrahydropyridine disappear. The decomposition process can be summarised as in eqn. (5.23).

$$\sim\!\!-N=P\!\!-\!\!\sim \xrightarrow{\Delta_1} \sim\!\!-N=\bar{P}\!\!-\!\!\sim + \bigcirc\!\!-\!\!N\!\!-\!\!H + \bigcirc\!\!-\!\!N \xrightarrow{\Delta_2} P_4 + NH_3 + N_2 + \bigcirc\!\!-\!\!N$$

(5.23)

The direct pyrolysis of poly(dimethylsiloxane) into the MS has been reported.[11] The mass spectrum of this polymer (otherwise known as OV-1

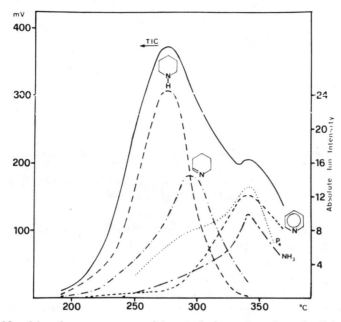

FIG. 5.23. Mass fragmentograms of the pyrolytic products for poly(dipiperidine-phosphazene).

stationary phase in gas chromatography) is shown in Fig. 5.24. Common fragments are derived from

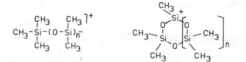

These data may be interpreted as indicating that there are two dominating thermal decomposition processes. The first is a cyclisation reaction occurring via an intramolecular exchange process (analogous to other polymers discussed above), while the second leads to the formation of open chain fragments through a methyl shift rearrangement involving the transfer of a methyl group from a neighbouring group to form a terminal trisylile group.

5.3.2. Structure Differentiation
Isomeric polymer structures can be, in general, well differentiated by MS.

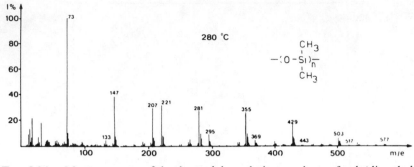

FIG. 5.24.   Mass spectrum of the thermal degradation products of poly(dimethyl-siloxane) at 280°C and 70 eV.

One notable exception is stereoisomeric polymers, since MS is relatively insensitive to certain types of stereoisomerism.

There have been attempts to detect cross-linked structures in polystyrene containing 8 % of pure *meta-* and *para*-divinylbenzene,[79,80] or to detect branching in polybenzyls,[65] but the differentiation is weak.

An interesting investigation of head-to-head (HH) and head-to-tail (HT) polystyrenes indicates their possible differentiation by MS.[62] In fact, their spectra (Fig. 5.25) present relevant differences. The mass spectrum of HT polystyrene, which decomposes thermally by depolymerisation, shows essentially the molecular ions of the monomer (m/e 104), dimer (m/e 208) and trimer (312). Instead the mass spectrum of HH polystyrene, which degrades by statistical cleavage into oligomeric styrenes, peaks up to mass 547 are found. The formation of stilbene (m/e 180) is a diagnostic reaction for HH polystyrene.

Another example of structure differentiation is given by four isomeric truxillic and truxinic polyamides.[43,44] GPy–EIMS of these polymers showed that the thermal degradation products are sensibly different for the head-to-tail and head-to-head isomers and that cyclobutane ring cleavage is predominant.

Accordingly, in the head-to-head isomers a stilbene molecule was evolved by thermal cleavage, whereas in the head-to-tail isomers the evolution of stilbene was not observed:

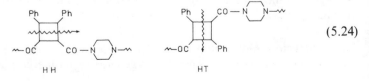

(5.24)

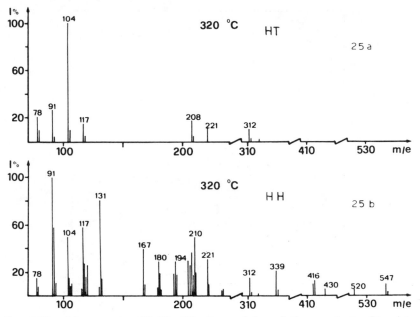

FIG. 5.25.   Mass spectra (18 eV) of the thermal degradation products of head-to-tail (HT) (24a) and head-to-head (HH) polystyrene (24b) at 320 °C.

Among the polycondensation polymers, isomeric polyesters and polyamides show well differentiated mass spectra.[33-35]

In another case, relevant differences are found in the mass spectra of *para*- and *meta*-poly(phenyleneoxide) although these have not been discussed by the authors.[18,55]

The examples reported show the great potential of MS in differentiating isomeric polymer structure. Further work in this direction is therefore worthwhile.

A systematic exploration of the possibility of structure differentiation of polymers by pyrolysis–MS should take into account that the latter depends also on the mechanisms of thermal decomposition.

5.3.3. Copolymer Analysis

5.3.3.1. Composition Analysis

The application of Py–MS to structure elucidation of copolymers has been performed in several instances. The composition of copolymers of α-methylstyrene with methylmethacrylate and acrylonitrile has been correlated with characteristic MS peak intensities.[63]

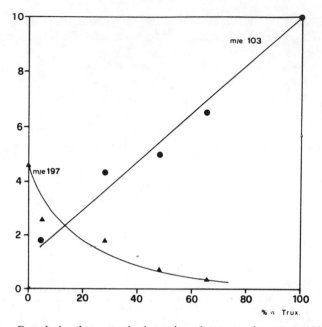

FIG. 5.26. Correlation between the intensity of mass peaks at m/e 103 and m/e 197, characteristic of the truxillic and adipic moieties, and the composition of the copolyamides. Abscissa, % α-Trux. Ordinate, I/I_{tot}.

The same has been done in the case of copolyesters from terephthalic and succinic acid with hydroquinone,[35] copolyesters from glycolic and lactic acid,[81] and copolyamides from adipic and truxillic acids with aliphatic amines.[82]

One example of these correlations, relative to the latter case, is shown in Fig. 5.26. As a general rule, it may be remarked that only mass peaks showing a linear correlation with the copolymer composition should be used to estimate the composition of unknown samples. Mass peaks showing a regular but non-linear correlation should be neglected because even a simple curved line indicates that spurious structures contribute to the intensity of these peaks, and therefore the accuracy of the composition estimation is drastically lowered.

5.3.3.2. Sequential Analysis

The alternating structure of some copolyamides from p-aminobenzoic and γ-aminobutyric acids has been proved by GPy–EIMS.[83]

In fact, the mass spectra of these polymers demonstrate that no pyrolysis product of fragment is formed containing neighbouring aromatic or aliphatic structure units.[83]

An elegant method of following the ester–ester exchange reactions of aliphatic polyesters by Py–EIMS, has recently been reported.[84]

In this work two homopolyesters, A and B, were allowed to react at 312 °C and the copolymers obtained were analysed by Py–EIMS, detecting the intensities of characteristic peaks related to A, B, and AB units (Fig. 5.27).

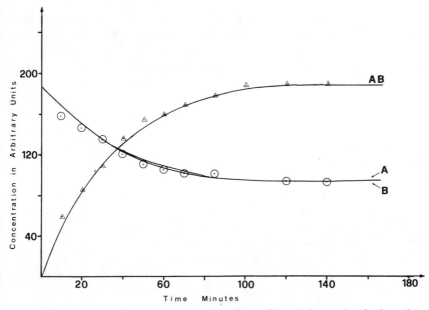

FIG. 5.27. Variation in the concentration of A and B and the randomised product AB for the exchange reaction at 312 °C.

The intensity of AB units was taken as the degree of randomisation of the copolymer formed. Since the MS spectra were taken at a pyrolytic temperature of 200 °C, there is little danger that additional thermal exchange reactions occur in the MS.

A detailed study to distinguish between blocked and random structures in low molecular weight copolymers of ethylene oxide and propylene oxide has recently been reported.[85] Using high-resolution MS analysis the authors have been able to measure approximate molecular weights, distribution of molecular species, and distribution of units in blocked and random copolymers.

5.3.4. Oligomers and Additives

5.3.4.1. Identification of Oligomers Contained in Polymeric Samples
Synthetic polymer samples frequently contain traces or sizeable amounts of monomer, solvents or low molecular weight oligomers.

Mass spectrometry is particularly suitable for the detection of these materials since they are volatile under high vacuum at relatively mild temperatures, while polymers stay undecomposed.

It is usual practice with the direct pyrolysis technique, to heat the polymer sample into the MS slightly below the temperature of thermal decomposition and to wait long enough to get a negligible TIC. After this stage is over, the linear programmed heating is started in order to get decomposition. Oligomeric species, if existing, are therefore seen in the pre-heating step. Several examples have been reported in the literature.[15,32,40,47,86]

A series of cyclic oligomers have been detected in poly-ε-caprolactam (nylon-6)[47] both by EI and FD techniques. Cyclic oligomers, which evaporate in the high vacuum of the ion source at temperatures below 200 °C, have been detected also in several other polyamides (nylon-7, nylon-12, nylon-4, nylon-6,6, nylon-6,10, nylon-4,10, nylon-11,6 and nylon-12,12).[40] Using the same technique, commercial poly(ethylene terephthalate) has been proved to contain about 1 % of the cyclic trimer.[86]

These examples illustrate that the direct MS technique can be routinely employed in order to detect oligomers eventually contained in polymer samples. It should be noted that the advantage of simplicity, sensitivity and high molecular weight detection capability, makes this technique highly preferable with respect to the conventional methods.

5.3.4.2. Analysis of Polymers Containing Additives
Volatile compounds added to polymers can be detected essentially with the same technique discussed in Section 5.3.4.1. Non-volatile additives may react with the polymeric substrate to generate volatile products that can be detected in the MS. Some inorganic additives, used as fire retardants, belong to this type. The systems most studied have been PVC/metal oxides[87,88] and PVC/PVB/metal oxides.[89] It has been found that the oxides of antimony, bismuth, zinc and tin form the corresponding chlorides or bromides, while those of iron, copper, molybdenum and aluminium do not form volatile compounds. The mass spectrum of a mixture PVC/PVB/Sb_2O_3 at 210 °C is shown in Fig. 5.28.

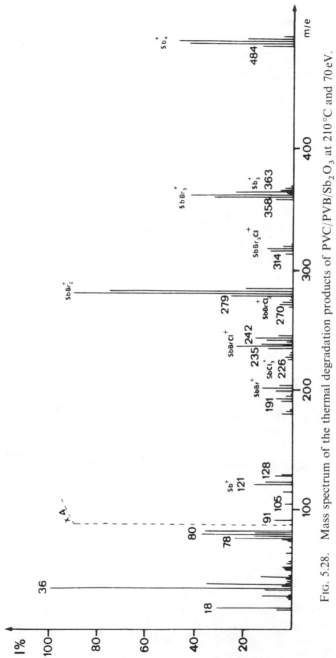

FIG. 5.28. Mass spectrum of the thermal degradation products of PVC/PVB/Sb$_2$O$_3$ at 210°C and 70 eV.

Peaks corresponding to the molecular ions of the four metal halides are easily detectable in the spectrum.

Also, the volatilisation curves of PVC appear to be significantly influenced by the addition of some metal oxides, and these results, combined with other information, have been used in order to study the mechanisms of action of these fire retardants.[67]

Mixtures of modacrylic fibres with Sb_2O_3 have also been studied by this technique.[70]

5.4. CONCLUSION

Considering the body of data reviewed in this chapter, one is forced to conclude that relevant efforts have been made in the last 15 years in order to develop MS of polymers into a mature field.

The high potential of MS with regard to structure elucidation of polymers and copolymers is becoming widely recognised and it becomes increasingly clear that the analysis of a polymeric sample by MS does not differ experimentally from the analysis of any low molecular weight (organic or inorganic) compound, and that it may be routinely performed into a commercial mass spectrometer.

Contrary to what might appear at first, the main difficulty in MS of polymers descends from the absence of a detailed theory of thermal degradation of polymers.

The development of such a theory has suffered in the past from a dramatic lack of information on the primary thermal fragmentation processes of polymers. As a consequence, the phenomenological approach provided by TGA and allied methods has largely been non-productive towards the elucidation of the mechanisms of thermal decomposition.

At present, without previous mechanistic knowledge, the structural information coming from Py–MS of an unknown polymer or copolymer is often ambiguous, so that its structure elucidation is difficult.

Most of the work presented in this review shows, however, that MS is a promising technique for investigating the thermal decomposition mechanisms of *known* polymers.

The next few years will tell us if the application of MS to the analysis of polymers and copolymers will produce significant progress in the understanding of polymers degradations.

In this case, polymer scientists will take full advantage of the extraordinarily abundant information provided by MS on polymer structure.

REFERENCES

1. *a* C. E. Roland Jones and C. A. Cramers (Eds.), *Analytical Pyrolysis*, Elsevier, Amsterdam, 1977. *b* M. J. Irwin, *J. of Analytical and Applied Pyrolysis*, 1, 3, 1979.
2. L. A. Wall, *J. Res. Nat. Bur. Stand.*, **41**, 315, 1948.
3. S. L. Madorsky and S. Straus, *J. Res. Nat. Bur. Stand.*, **40**, 417, 1948.
4. P. D. Zemany, *Anal. Chem.*, **24**, 1707, 1952.
5. S. L. Madorsky, *Thermal Degradation of Organic Polymers*, Interscience, NY, 1964, and references therein.
6. P. Bradt, V. H. Diebler and F. H. Mohler, *J. Res. Nat. Bur. Stand.*, **50**, 201, 1953; **55**, 323, 1955.
7. H. L. Friedman, in *High Temperature Polymers*, C. H. Segal (Ed.), Marcel Dekker, Inc., NY, 1967, p. 57, and references therein.
8. G. P. Shulman, in *High Temperature Polymers*, C. H. Segal (Ed.), Marcel Dekker, Inc., NY, 1967, p. 107, and references therein.
9. H. G. Langer and R. S. Gohlke, *Anal. Chem.*, **35**, 1301, 1963.
10. H. D. R. Schüddemagge and D. O. Hummel, *Advan. Mass Spectr.*, **4**, 857, Institute of Petroleum, London, 1968.
11. A. Zeman, *Angew. Makromol. Chem.*, **31**, 1, 1973.
12. D. O. Hummel, in *Analytical Pyrolysis*, C. E. Roland Jones and C. A. Cramers (Eds.), Elsevier, Amsterdam, 1977, p. 117.
13. R. D. Sedgwick, in *Developments in Polymer Characterisation—1*, J. D. Dawkins (Ed.), Applied Science Publishers Ltd, London, 1978, p. 41.
14. I. Lüderwald, in *Proceedings of the 5th European Symposium on Polymer Spectroscopy*, Cologne, September 1978, Verlag Chemice, Weinheim, 1979, p. 217.
15. R. H. Wiley, *Macromolecular Reviews*, **14**, 379, 1979.
16. H. R. Schulten and W. Görtz, *Anal. Chem.*, **50**, 428, 1978.
17. R. M. Lum, *J. Polym. Sci., Polym. Chem. Ed.*, **15**, 489, 1977.
18. D. O. Hummel, H. J. Düssel, H. Rosen and K. Rübenacker, *Makromol. Chem., Suppl. 1*, 471, 1975.
19. I. Lüderwald, M. Przybylski and H. Ringsdorf, in *Mass Spectrometry in Biochemistry and Medicine*, A. Frigerio and N. Castagnoli (Eds.), Raven Press, NY, 1974, p. 245.
20. H. S. Schulten, in *Methods of Biochemical Analysis*, **24**, J. Wiley and Sons, NY, 1977.
21. Y. Shimizu and B. Munson, *J. Polym. Sci., Polym. Chem. Ed.*, **17**, 1991, 1979.
22. T. Matsuo, H. Matsuda and I. Katakuse, *Anal. Chem.*, **51**, 1329, 1979.
23. K. Levsen and H. R. Schulten, *Biomedical Mass Spectrom.*, **3**, 137, 1976.
24. A. Ballistreri, S. Foti, P. Maravigna, G. Montaudo and E. Scamporrino, *J. Polym. Sci., Polym. Chem. Ed.*, **18**, 1923, 1980.
25. H. L. Friedman, in *High-temperature Polymers*, C. L. Segal (Ed.), Marcel Dekker, Inc., NY, 1967, p. 57.
26. H. L. Friedman, *Thermochim. Acta*, **1**, 199, 1970.
27. G. P. Shulman, in *High-temperature Polymers*, C. L. Segal (Ed.), Marcel Dekker, Inc., NY, 1967, p. 107.

28. *a*. K. H. Van Heek, H. Juentgen and W. Peters, *Ber. Bunsenges. Phys. Chem.*, **71**, 113, 1967. *b*. A. L. Yergey, F. W. Lampe, M. L. Vestal, A. G. Day, G. J. Fergusson, W. H. Johnston, J. S. Snyderman, R. H. Essenligh and J. E. Hudson, *Ind. Eng. Chem. Process Des. Develop.*, **13**, 233, 1974.
29. T. H. Risby and A. L. Yergey, *Anal. Chem.*, **50**, 327, 1978.
30. S. Foti, G. Montaudo and K. Müller, to be published.
31. I. Lüderwald, in *Developments in Polymer Degradation—2*, N. Grassie (Ed.), Applied Science Publishers Ltd., London, 1979, p. 77.
32. I. Lüderwald and H. Urrutia, in *Analytical Pyrolysis*, C. E. Roland Jones and C. A. Cramers (Eds.), Elsevier, Amsterdam, 1977, p. 117.
33. I. Lüderwald and H. Urrutia, *Makromol. Chem.*, **177**, 2093, 1976.
34. I. Lüderwald and H. Urrutia, *Makromol. Chem.*, **177**, 2079, 1976.
35. C. Aquilera and I. Lüderwald, *Makromol. Chem.*, **179**, 2817, 1978.
36. E. Jacobi, I. Lüderwald and R. C. Schulz, *Makromol. Chem.*, **179**, 429, 1978.
37. A. R. Kricheldorf and I. Lüderwald, *Makromol. Chem.*, **179**, 421, 1978.
38. R. A. Wiley, *J. Macromol. Sci.-Chem.*, **A4**(8), 1797, 1970.
39. I. Lüderwald, *Makromol. Chem.*, **178**, 2603, 1977.
40. I. Lüderwald and F. Merz, *Angew. Makromol. Chem.*, **74**, 165, 1978.
41. H. J. Düssel, H. Rosen and D. O. Hummel, *Makromol. Chem.*, **177**, 2343, 1976.
42. S. Foti, I. Lüderwald, G. Montaudo and M. Przybylski, *Angew. Makromol. Chem.*, **62**, 215, 1977.
43. S. Caccamese, P. Maravigna, G. Montaudo and N. Przybylski, *J. Polym. Sci., Polym. Chem. Ed.*, **13**, 2061, 1975.
44. I. Lüderwald, M. Przybylski, H. Ringsdorf, S. Foti and C. Montaudo, in *Analytical Pyrolysis*, C. E. Roland Jones and C. A. Cramers (Eds.), Elsevier, Amsterdam, 1977, p. 297.
45. I. Lüderwald and H. Ringsdorf, *Angew. Makromol. Chem.*, **29/30**, 441, 1973.
46. I. Lüderwald and H. Ringsdorf, *Angew. Makromol. Chem.*, **29/30**, 453, 1973.
47. I. Lüderwald, F. Merz and M. Rothe, *Angew. Makromol. Chem.*, **67**, 193, 1978.
48. A. Ballistreri, S. Foti, P. Maravigna, G. Montaudo and E. Scamporrino, *Makromol. Chem.*, **181**, 2161, 1980.
49. S. Foti, P. Maravigna and G. Montaudo, *J. Polym. Sci., Polym. Chem. Ed.*, **19**, 1679, 1981.
50. S. Caruso, S. Foti, P. Maravigna and G. Montaudo, *J. Polym. Sci., Polym. Chem. Ed.*, in press.
51. A. Davis and J. H. Golden, *J. Macromol. Sci. Rev. Macromol. Chem.*, **C3**, 49, 1969.
52. R. H. Wiley, *Macromolecules*, **4**, 254, 1971.
53. S. Foti and G. Montaudo, unpublished results.
54. G. Montaudo, M. Przybylski and H. Ringsdorf, *Makromol. Chem.*, **176**, 1753, 1975.
55. G. Montaudo, M. Przybylski and H. Ringsdorf, *Makromol. Chem.*, **176**, 1763, 1975.
56. R. H. Wiley, *J. Polym. Sci.*, **A-1**, **9**, 129, 1971.
57. G. Bruno, S. Foti, P. Maravigna, G. Montaudo and M. Przybylski, *Polymer*, **18**, 1149, 1977.
58. F. Bottino, S. Foti, G. Montaudo, S. Pappalardo, I. Lüderwald and M. Przybylski, *J. Polym. Sci., Polym. Chem. Ed.*, **16**, 3131, 1978.

59. F. Bottino, S. Foti, G. Montaudo, S. Pappalardo, I. Lüderwald and M. Przybylski, *Angew. Makromol. Chem.*, **67**, 203, 1978.
60. P. G. Kistemaker, A. J. H. Boerboom, H. L. C. Meuzelaar, *Dynamic Mass Spectrometry*, Vol. 4, Ch. 9.
61. D. O. Hummel, H. J. Düssel and K. Rübenacker, *Makromol. Chem.*, **145**, 267, 1971.
62. I. Lüderwald and O. Vogl, *Makromol. Chem.*, **180**, 2295, 1979.
63. D. O. Hummel and H. J. Düssel, *Makromol. Chem.*, **175**, 655, 1974.
64. I. Lüderwald, G. Montaudo, M. Przybylski and H. Ringsdorf, *Makromol. Chem.*, **175**, 2423, 1974.
65. R. W. Lenz, I. Lüderwald, G. Montaudo, M. Przybylski and H. Ringsdorf, *Makromol. Chem.*, **175**, 2441, 1974.
66. A. Ballistreri, S. Foti, G. Montaudo and E. Scamporrino, *J. Polym. Sci., Polym. Chem. Ed.*, **18**, 1147, 1980.
67. A. Ballistreri, S. Foti, P. Maravigna, G. Montaudo and E. Scamporrino, *J. Polym. Sci., Polym. Chem. Ed.*, **18**, 3101, 1980.
68. A. Ballistreri, S. Foti, P. Maravigna, G. Montaudo and E. Scamporrino, *Polymer*, **22**, 131, 1980.
69. A. Ballistreri, S. Foti, G. Montaudo, S. Pappalardo, E. Scamporrino, A. Arnesano and S. Calgari, *Makromol. Chem.*, **180**, 2835, 1979.
70. A. Ballistreri, S. Foti, G. Montaudo, S. Pappalardo, E. Scamporrino, A. Arnesano and S. Calgari, *Makromol. Chem.*, **180**, 2843, 1979.
71. M. Przybylski, H. Ringsdorf and H. Ritter, *Makromol. Chem., Suppl. 1*, 297, 1975.
72. I. Lüderwald, M. Przybylski and H. Ringsdorf, in *Advances in Mass Spectrometry in Biochemistry and Medicine*, Vol. II, Spectrum Publications, 1976.
73. G. A. Chanock and J. L. Loo, *Anal. Biochem.*, **27**, 81, 1970.
74. H. R. Schulten, H. D. Beckey, A. J. Boerboom and H. L. C. Meuzelaar, *Anal. Chem.*, **45**, 2358, 1973.
75. M. A. Posthumus, N. M. M. Nibbering, A. J. H. Boerboom and H. R. Schulten, *Biomed. Mass Spectrom.*, **1**, 352, 1974.
76. H. R. Schulten, in *New Approaches to the Identification of Microorganisms*, Part A: *New Technologies in the Automation of Microbiological Identification Routines*, J. Wiley & Son, NY, 1975, p. 155.
77. H. R. Schulten, in *Analytical Pyrolysis*, C. E. Roland Jones and C. A. Cramers (Eds.), Elsevier, Amsterdam 1977, p. 17.
78. A. Ballistreri, S. Foti, G. Montaudo, S. Lora and G. Pezzin, *Makromol. Chem.*, **182**, 1319, 1981.
79. R. H. Wiley, *J. Polym. Sci., Part A-1*, **8**, 792, 1970.
80. R. H. Wiley and L. H. Smithson, *J. Macromol. Sci., Chem.*, A2, 589, 1968.
81. E. Jacobi, I. Lüderwald and R. C. Schulz, *Makromol. Chem.*, **179**, 277, 1978.
82. S. Caccamese, S. Foti, P. Maravigna, G. Montaudo, A. Recca, I. Lüderwald and M. Przybylski, *J. Polym. Sci., Polym. Chem. Ed.*, **15**, 5, 1977.
83. I. Lüderwald, *Angew. Makromol. Chem.*, **56**, 173, 1976.
84. H. G. Ramjit and R. D. Sedgwick, *J. Macromol. Sci., Chem.*, **A10**, 815, 1976.
85. A. K. Lee and R. D. Sedgwick, *J. Polym. Sci., Polym. Chem. Ed.*, **16**, 685, 1978.

86. I. Lüderwald, H. Urrutia, H. Herlinger and P. Hirt, *Angew. Makromol. Chem.*, **50**, 163, 1976.
87. A. Ballistreri, S. Foti, G. Montaudo, S. Pappalardo and E. Scamporrino, *Chimica e Industria*, **60**, 501, 1978.
88. A. Ballistreri, S. Foti, G. Montaudo, S. Pappalardo and E. Scamporrino, *J. Polym. Sci., Polym. Chem. Ed.*, **17**, 2469, 1979.
89. A. Ballistreri, S. Foti, G. Montaudo, S. Pappalardo and E. Scamporrino, *Polymer*, **20**, 783, 1979.

Chapter 6

THERMAL METHODS OF ANALYSIS OF POLYMERS

JAMES N. HAY

Department of Chemistry, University of Birmingham, UK

6.1. INTRODUCTION

Thermoanalytical techniques are being increasingly used to characterise a polymer, measure the temperature dependence of some mechanical or physical property and correlate it to structure, so much so that the number of new techniques and their applications makes it very difficult for the average materials scientist to keep abreast of developments except in a limiting number of areas. A daunting task faces anyone who attempts to give anything other than a superficial review of these. Fortunately the development of the subject has been well reviewed in the past and free use of these will be made.[1-14] So much effort has been invested in these techniques that of the 10 000 analytical units recently reported[1] as being used world-wide, the majority are routinely being devoted to solving polymer problems of a wide range of diversity. This activity derives in part from an appreciation that the thermal history, heat treatment and fabrication conditions are important in determining the ultimate properties and working conditions of polymers, and also that these can be measured simply and routinely in thermal analytical techniques. This appreciation has been paralleled with a massive investment in commercial equipment with a proliferation of new techniques and improved procedures. One recent innovation, which will undoubtedly continue these improvements in reproducibility and in the quality of the data,[15] is microprocessor control. This alone will ensure that the activity and developments of the past decade will continue at an accelerated pace.

Nevertheless, despite this activity there still remains considerable doubt about the viability of such procedures to measure meaningful parameters characteristic of the polymer and independent of the measuring procedure, and the literature abounds with warnings about their applicability to polymer systems,[3] listing the effect of experimental and instrumental variables. Measurements are, of course, made under non-equilibrium conditions, on ill defined, highly viscous, poly-dispersed insulators, such as polymers! In the light of such comments, what value can be placed on the data obtained by such means, and what limitations are inherent to each technique?

6.2. RANGE OF TECHNIQUES

Thermal analyses are extremely diverse, embracing a large number of very different techniques (see Table 6.1) where the techniques recognised by ICTA are reproduced.[1] They involve 'a group of techniques in which

TABLE 6.1
Techniques Recognised by ICTA[a]

Abbreviations approved	Property	Technique
TG	Mass	Thermogravimetry
		Isobaric Mass Change
EGA	Mass	Evolved Gas Analysis
EGD	Mass	Evolved Gas Detection
		Thermoparticulate Analysis
DTA	Temperature	Differential Thermal Analysis
		(Heating/cooling curves)
DSC[b]	Enthalpy	Differential Scanning Calorimetry
	Length	Thermodilatometry
TMA	Modulus	Thermomechanical Analysis
DTM	Modulus	Dynamic Thermomechanical Analysis

[a] Other techniques include sound and light transmission and emission, electrical and magnetic properties.
[b] Both power compensation and heat flux DSC are recognised.

properties are measured as a function of temperature, or time if the temperature is constant, while the substance under examination is subjected to a controlled temperature programme'. This makes the subject of thermal analysis extremely broad since, in theory at least, there is no restriction on the properties measured. For the present purposes, however, a limit will be imposed to those properties which can be measured simply without resort to a complex spectrometer, or other equipment and excluding those techniques which require much more equipment to measure the property than control the temperature. Inevitably such a personal definition is arbitrary, but the essential spirit of it is that while the temperature dependence of the i.r., NMR and X-ray scattering spectra, for example, is most useful in structural determination, the temperature control of the sample is trivial compared with the measurement of the spectra, and is not a thermo-analytical technique in essence but is part of the spectroscopy. It is not advocated that these measurements should not be made, but to indicate why they have been excluded from the present discussion. This restricts the techniques to the measurement of mass, volume, temperature, heat flux, etc.

The ICTA recognised list of thermo-analytical techniques,[1] with their accepted abbreviations, has been reproduced in Table 6.1. Other derivative techniques not listed, involving the rate of change of the property with time and temperature, are also supplied. Multiple techniques involving the simultaneous measurement of two or more properties, e.g. mass and temperature, have also been adopted widely. These will not be considered since they can be regarded as the sum of the individual components, and invariably the two techniques do not, of necessity, have the same optimum working conditions, and the best results cannot be achieved with both in a single determination.

The general applicability of the various techniques to polymers will be considered in detail under each separate technique, along with their associated problems and the effect of procedural variables. It should, however, be appreciated that the diverse nature of thermal analyses of polymers makes the subject diffuse and the literature difficult to assimilate. Research papers are widely scattered over technological and pure research journals in subjects from theoretical physics to applied chemistry. This diversity of approach and interest makes it essential that the same standard of reporting thermo-analytical data be universally adopted, with sufficient details on standardisation and calibration that the experiments can be reproduced by others. The procedures adopted by the ICTA's standardising committee, and subsequently adopted internationally, should be followed as far as possible.[1] The information required varies with each technique and will be considered separately under each technique.

Finally, each technique has been extensively reviewed, and there are excellent instrument manuals and report sheets published by the manufacturers which will guide the uninitiated through the pitfalls of each technique and which are essential reading.

6.3. TEMPERATURE DEPENDENCE OF MASS

Techniques arising from measurement of the changes in weight, or the production of volatile substances, with increase in the temperature of the sample, are widely used to characterise the molecular composition, and have been applied to a wide range of polymeric studies, see Table 6.2. There are similar problems associated with the application of each of these techniques and it will only be necessary to deal with a few to appreciate their nature.

TABLE 6.2
Applications of Mass Measuring Techniques

Degradation profiles used for 'fingerprint' analyses.
Quality control to determine effect of additives.
Thermal stability in inert atmosphere.
Oxidative stability in air, or oxygen-rich atmosphere (flammability and fire-retardancy).
Adsorption–desorption studies.
Kinetics of mass evolution—degradation, analysis of products, kinetics of products evolution.

6.3.1. Thermogravimetry (TG)
As the name implies, this involves monitoring continuously the weight of a sample with temperature at a constant rate of change of temperature or with time isothermally. These studies are carried out in a static or a flowing atmosphere of an active or inert gas. All these points, i.e. weight, rate of heating, temperature measurement, atmosphere, etc., introduce variables. The effect of these on the measured weight/temperature plots should be determined and allowances made for thermal lag between thermal sensors and sample. All details should be listed in the report in such a way that others can repeat the determinations.

Derivative thermogravimetric analysis (DTG) monitors the rate of change of weight with time plotted against temperature and is particularly useful in defining temperatures of initial onset of decomposition and maximum rates of decomposition. The presence of competing reactions can be observed more readily than from the initial TG curves. However, it should be appreciated that these are dynamic tests and rate of heating displaces these temperatures. Nevertheless, for an identical set of conditions, similar polymer systems can effectively be compared (see Fig. 6.1 in

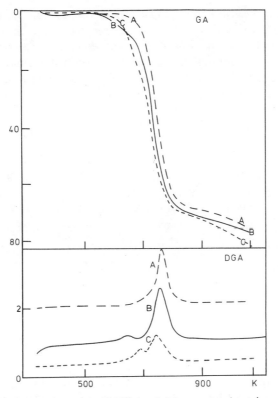

Fig. 6.1.   Thermogravimetric and differential thermogravimetric analysis of poly-(bis-phenol-A terephthalate) (A) and block copolycarbonates (B and C). Weight loss and weight loss rate against temperature. Nitrogen gas flow rate, 2 cm³ min⁻¹. Heating rate, 2 K min⁻¹. Sample weight 2 mg, as a finely divided powder. After C.P. Bosnyak.[16]

which the thermal stability of poly(bis-phenol-A-terephthalates) are compared with their copolycarbonates). This also exhibits well the difference between TG and DTG. Similarities in decomposition characteristics imply that the common bis-phenol-A-terephthalate units rather than the carbonate are important in limiting their thermal stability. Further rate studies confirm these conclusions.[16]

Thermogravimetry has been used widely to study all physical processes involving weight changes and as such has been used to measure diffusion characteristics into polymers, and desorption characteristics. Moisture uptake, in particular, has been well studied since hydrolysis of esters and carbonates can limit the moulding temperature adopted. It is normal to define a drying procedure which reduces substantially the amounts of water present. However, by far the commonest use for TG is in measuring the thermal and oxidative stability of polymers under working conditions. Since all organic molecules have a limiting thermal stability and will decompose either completely or to a limiting extent, to a more or less stable residue, the technique is applicable to all polymers. These TG or DTG decomposition curves give a 'fingerprint' method of classifying polymer and copolymer systems. Such procedures require an extensive library of decomposition curves of well characterised polymeric systems assimilated in previous studies. These 'fingerprint' spectra vary substantially with experimental variables, such as sample size and form, heating rate and atmosphere. Defining a standard test for a particular purpose is readily done, but it should be appreciated that these are relative and not absolute. Certainly similar trends in thermal stability have been observed[2] within a series of polymers using very dissimilar experimental conditions. Experimental and instrumental variables[2] have a pronounced effect and these should be listed in repeating such studies.

Conditions, however, can be selected,[16] e.g. slow heating rates, low sample weight and finely divided powders, such that they have little effect on the overall decomposition curves. Under these conditions, the stability has significance in terms of molecular parameters, independent of experimental and instrumental variables. Even under these conditions, however, MacCallum[17] has pointed out the severe limitations imposed on mechanistic studies based on TG derived rate curves.

The most general rate expression for a bulk decomposition is

$$-\,\mathrm{d}N/\mathrm{d}t = k(N)^n/(V)^{n-1} \tag{6.1}$$

in which k is the absolute rate constant, n the external reaction order, N the number of moles of reactant producing volatiles, and V the volume of the reacting system.

Modifying this equation for TG, the weight change is then

$$-\,\mathrm{d}w/\mathrm{d}t = k(\rho/m)^{n-1}w \tag{6.2}$$

in which ρ is the density of the melt, and m the molar mass of the reacting units. Accordingly, it is inherent in the nature of TG that rate studies based on weight of volatilised materials always gives an apparently first-order reaction irrespective of the external order n, and an incorrect measure of the absolute rate constant, k, unless of course the order is exactly unity.

Since the temperature dependence of the rate constant is given by the Arrhenius expression,

$$k = A \exp - \Delta E / kT \qquad (6.3)$$

then estimates of the pre-exponential factor, A, from the plot of log (k_{obs}) against $1/T$, will be incorrect, although ΔE will be meaningful provided there is no change in n value with temperature. Accordingly, mechanistic studies based on the value of the pre-exponential factor A, and the entropy of formation of the transition state cannot be seriously entertained.

MacCallum[17] further warns that this is the simplest possible situation in which polymer depolymerises to monomer, and the situation would become increasingly complex if residues are present as products of the decomposition, or if the initial undegraded polymer was partially crystalline. Further information is required about the decomposition of various molecular weight species if a mechanism is to be proposed. Indeed, Barlow et al.[18] have outlined detailed mechanistic schemes which lead to a different dependence of the observed first-order rate constant, $k_{obs} \equiv k(\rho/m)^{n-1}$, on the number average degree of polymerisation, x_n, i.e.

(a) Decomposition initiated by random scission of the chain followed by depolymerisation without transfer or termination to the end of the chain, then $k_{obs} \propto x_n$.
(b) Decomposition initiated at the end units of the chain followed by depolymerisation to the end of the chain, then k_{obs} is independent of x_n.
(c) Decomposition initiated at the end units of the chain with first-order termination, then $k_d \propto x_n^{-1}$
(d) Decomposition initiated at the end units of the chain with second-order termination, then $k_d \propto x_n^{-1/2}$

Further information on the mechanism can, of course, be determined by a study of the products of the decomposition as a function of conversion, but this involves other techniques.

6.3.2. Gas Evolution Studies

Decomposition with evolution of volatiles into a continuously evacuated system exerts a small pressure characteristic of the rate of production of the volatiles. This method was adopted by Grassie and Melville[19] in their dynamic molecular still, and has been widely used by a number of workers in polymer degradation studies.[14] McNeill[20] has modified the technique by incorporating programmed heating and produced the term 'thermal

volatilisation analysis'. Further modifications included differential condensation of the products according to their volatilities by condensing them
into traps of different temperatures in an attempt to measure simultaneously the concentrations of each component during the decomposition. Where these techniques have been extensively used to determine
kinetic parameters they suffer the same constraints as defined above by
MacCallum.[17] However, by systematically altering the degrading system
by incorporating model structures, weak links, etc., the techniques have
been successfully applied to mechanistic studies.[14,20]

A particularly useful technique for polymer degradation studies is
pyrolysis gas liquid chromatography,[21] especially as extended to micro
samples.[22] Programme controlled heating eliminates over-heating of the
micro-specimens, and a voltage compensator reduces the heating from a
'booster' heater proportional to the thermocouple reading of the oven
temperature. This reduced heat-up times to 20–100 ms and enabled
isothermal rate studies to be made on samples pyrolysed for 1–30 s with no
broadening of the GLC peaks on analysis.

The technique is particularly useful in that the individual volatile
components can be monitored separately. By sweeping them into the
columns for analysis, there is little opportunity for the products to react
with one another and so the primary products of decomposition alone are
being determined. The individual rate constants for production of each
component can then be separately determined and activation energies
established. Competing reactions can thus be established in the degradation. Obviously this is of considerable advantage when many
components are produced, as in the case of polyacrylonitrile[23–25] which
above 300 °C evolves ammonia, hydrogen cyanide, acrylonitrile, acetonit-
rile and methacrylonitrile in a series of competing reactions. This can
readily be seen from a plot of relative yield of each component as a function
of decomposition temperature (see Fig. 6.2).

The thermal depolymerisation of polystyrenes prepared by anionic and
free radical initiations has been studied recently by pyrolysis GLC.[26] This
has been studied by other techniques: TG[27,28] and gas evolution detection,[29] and it is generally accepted that depolymerisation proceeds by a
radical chain mechanism in which monomer is the main product. Random
scission occurs initially or during depolymerisation and the molecular
weight falls rapidly, but the nature of the initiation sites, whether these are
randomly placed weak links or end-groups, is controversial. MacCallum[33]
has reviewed the conflicting evidence.

Lehrle and Peakman[26] confirmed that the dominant species in the

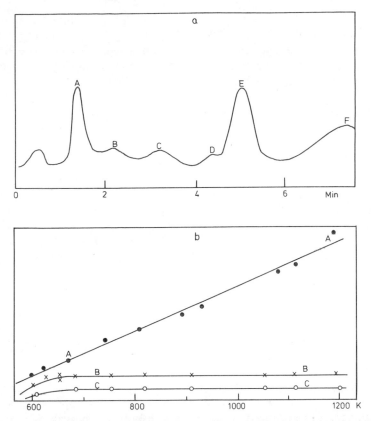

FIG. 6.2. Polyacrylonitrile degradation. (a) Products determined by GLC, 0·01 mg sample degraded at 625 K for 30 s. GLC columns coated with tricresyl-phosphate. Relative intensities plotted against retention time. A, ammonia; B, not determined; C, hydrogen cyanide; D, methacrylonitrile; E, acrylonitrile; F, vinylacetonitrile. (b) Dependence of the limiting yield of the volatile products on temperature. A, acrylonitrile; B, methacrylonitrile; C, hydrogen cyanide. After F. A. Bell *et al.*[23]

depolymerisation was monomer, with small amounts of dimer, trimer, etc. The yield of oligomers did not vary from sample to sample or with decomposition temperature, and indeed oligomers were not present on degrading very thin films. The observed first-order rate constant for monomer production decreased with molecular weight, suggesting that the predominant mechanism involved chain end initiation, but the mode of termination was uncertain since plots of rate constants against either

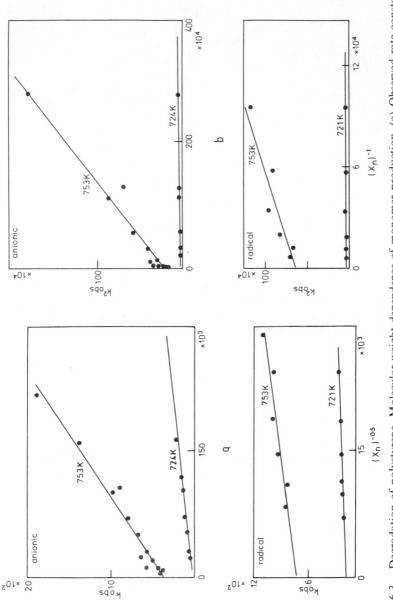

Fig. 6.3. Degradation of polystyrene. Molecular weight dependence of monomer production. (a) Observed rate constant (k_{obs}) dependence on the square root of the reciprocal degree of polymerisation (X_n); free radical and anionic polymerised polymers. (b) Observed rate constants (k_{obs}) squared dependence on the reciprocal degree of polymerisation (X_n). After R. E. Peakman.[26]

reciprocal degree of polymerisation or the square root were equally linear, Fig. 6.3.

Since, for the two mechanisms, i.e. (c) and (d) above, end initiation is given by:

$$P \longrightarrow R^{\cdot} + x^{\cdot}$$

where P is polymer chain concentration and $R^{\cdot}$ the radical. Then, the initiation rate, R_i, is:

$$R_i = k_{i,e}P$$

where $k_{i,e}$ is the initiation rate constant.

Random scission initiation is given by:

$$\sim\!\!C\!\!-\!\!C\!\!\sim \longrightarrow 2R^{\cdot}$$

where the initiation rate, R_i', is:

$$R_i' = k_{i,s}P$$

Depropagation

$$R_n^{\cdot} \longrightarrow R_{n-1}^{\cdot} + M$$

and rate of monomer production is

$$R_D = k_d[R^{\cdot}]$$

where k_d is the depolymerisation rate constant.

Termination (a) unimolecular is:

$$R^{\cdot} \longrightarrow \text{Dead polymer}$$

then rate of termination, R_t, is:

$$R_t = k_{u,t}[R^{\cdot}]$$

and, (b) bimolecular is:

$$R^{\cdot} + R^{\cdot} \longrightarrow P$$

and the rate of termination, R_t', is:

$$R_t' = k_{b,t}[R^{\cdot}]^2$$

Unimolecular termination could be considered as termination of geminate radicals created in one act of initiation, since this will be controlled by diffusion, or to be transfer to monomer with loss of the volatile radical.

For the two termination processes and assuming a stationary concentration of radicals, the above scheme leads to the overall rate of production of monomer. For unimolecular termination,

$$\frac{d}{dt}[n] = \frac{k_{i,e}}{x_n} + 2k_{i,s}\frac{k_d}{k_{u,t}}(w_o - w) \tag{6.4}$$

where w_o is the initial mass of polymer and w is the mass evolved.

The observed first-order rate constant for monomer production is:

$$k_{obs} = \frac{A}{x_n} + B$$

with

$$A = \frac{k_{i,e}k_d}{k_{u,t}}$$

and

$$B = \frac{2k_{i,s}k_d}{k_{u,t}}$$

For bimolecular termination,

$$\frac{d}{dt}[n] = \frac{k_{i,e}}{x_n} + 2k_{i,s}^{\frac{1}{2}}\left(\frac{k_d\mu^{1/2}}{2k_{b,t}\rho}\right)^{1/2}(w_o - w) \tag{6.5}$$

and the observed first-order rate constant, k_{obs}, is such that:

$$(k_{obs})^2 = \frac{A'}{x_n} + B'$$

with

$$A' = \frac{k_{i,e}k_d^2\mu}{2k_{b,t}\rho}$$

and

$$B' = \frac{k_{i,s}k_d^2\mu}{k_{b,t}\rho}$$

in which ρ is the bulk density and μ the monomer molecular weight.

According to the termination mechanism, plots of k_{obs} or $(k_{obs})^2$ against the reciprocal degree of polymerisation will be linear. However, as shown in Fig. 6.3 no distinction could be made by this method between the two mechanisms. However, since the constants, A, A', B and B' (determined from slopes and intercepts) are ratios of rate constant, then their Arrhenius dependence corresponds to activation energy differences, in particular

$$E_{A'} = E_{i,e} + 2E_d - E_{b,t} \quad \text{and} \quad E_{B'} = E_{i,s} + 2E_d - E_{b,t} \quad (6.6)$$

and

$$E_A = E_{i,e} + E_d - E_{u,t} \quad \text{and} \quad E_B = E_{i,s} + E_d - E_{u,t} \quad (6.7)$$

Taking $E_d = E_p - \Delta H_p$ (calculated) where E_p is activation energy for polymerisation of styrene, ΔH_p (calculated) the heat of polymerisation $(-4\cdot47\,\text{kJ mol}^{-1})$ and $E_t \simeq 0\,\text{kJ mol}^{-1}$, then the activation energies for the initiation steps, listed in Table 6.3, were obtained.

TABLE 6.3
Activation Energies for Initiation (kJ mol^{-1})

Sample	Random scission	End initiation
(a) Unimolecular Termination		
Radical	249 ± 16	146 ± 34
Ionic	334 ± 11	249 ± 16
(b) Bimolecular Termination		
Radical	538 ± 24	415 ± 56
Ionic	744 ± 79	558 ± 14

E_p was taken to be $1\cdot87\,\text{kJ mol}^{-1}$.

Since carbon–carbon aliphatic bonds have dissociation energies less than 300–$400\,\text{kJ mol}^{-1}$ the bimolecular termination mechanism predicts values very much in excess of this value and so is implausible. The unimolecular termination mechanism predicts values which are considerably less, e.g. $330 \pm 10\,\text{kJ mol}^{-1}$ for anionic polystyrene which is pleasingly close to a value of $370\,\text{kJ mol}^{-1}$ for the aliphatic chain bonds. End initiation must involve cleavage of C—H bond, or Bu—C bond, but since dissociation energies are 450 and $290\,\text{kJ mol}^{-1}$, respectively, this would suggest that the butyl group, obtained from the alkyl lithium initiator, cleaves in the initiation step.

Free radical prepared polystyrenes are much more labile, and lower initiation energies are involved consistent with the predictions of weak link scission based on molecular weight studies on the degraded polymer.[27,29,30–32] The random scission initiation's activation energy is consistent with bond dissociation energy of a monomer head-to-head link, of a branched point on the chain, and of bonds adjacent to unsaturated group produced by polymerising through the ring positions. The extremely low values for end-group initiation can only be reconciled by an unusual structure, e.g. peroxide, etc. Unfortunately, the nature of the end-groups and the free radical initiator are unknown.

This is a fine example of the strength of pyrolysis-GLC in tackling polymer degradation studies. Since small amounts of materials are pyrolysed in each determination, studies can be made on fractionated materials for which only small amounts are available, particularly those obtained from GPC.[34] In this way molecular weight effects can be explored thoroughly.

6.3.3. Thermoparticulate Analysis

This technique was first applied by Doyle[35] to detect polymer decomposition and uses the detection of condensation nuclei evolved during decomposition. It is particularly useful in detecting the onset of decomposition at lower extents and lower temperatures than previously determined.

Two analytical instruments are available commercially for the detection of airborne particles[36] which include ion-chamber detectors and a cloud chamber in which water vapour condenses around the particles.[37] The change in current from the ion-chamber is proportional to the concentration of particulate matter flowing through the chamber, and the water droplets scatter light, the amount of scattering being proportional to the number of nuclei. However, both detectors measure different size particles—particles above 20–30 Å in diameter appear to be detected by the ion-chamber while Aitken particles $(100–10^4 \text{ Å})$ act as nuclei for condensing water.

The technique has recently been extended by Smith et al.[38] to the decomposition of model low molecular weight organic compounds from which it would appear that oligomeric clusters of the molecules are being detected. By analogy thermoparticulate analysis of polymers must be detecting large chain fragments driven off on heating. The importance of this technique in polymer degradation and polymer analysis still remains to be assessed.

Evolved gas, such as hydrogen chloride, which does not alone cause condensation or ion-chamber detection, can be studied by reacting to form ionic solids, e.g. ammonia,[39] which generate nuclei. This analysis is, of course, specific to the one product of the decomposition, and it can be estimated in the presence of many others.

An analogous technique has been used by Dillon *et al.*[40] using a modified time-of-flight mass spectrometer to study the emission of positive ions from various polymers bombarded by ions with variable kinetic energies. Polymers were heated from 300 to 600 K and the products analysed, as they were formed in the mass spectrometer, to determine their mass and kinetic energy distribution. Large fragments characteristic of the monomeric units of the polymer were observed, their size being determined by the kinetic energy of the bombarding ions.

The technique has the potential for determining the structure and the composition of polymer surfaces and in particular analysing surface contamination, adsorption and oxidation. Furthermore, with the introduction of etching by ion-beams in semi-conductor manufacture, this technique should be able to monitor the processes involved.

6.4. THERMAL ANALYSIS

6.4.1. Differential Thermal Analysis and Differential Scanning Calorimetry

Differential thermal analysis (DTA) involves the measurement of temperature differences between sample and reference as a function of time or temperature (see Fig. 6.4) as heat is supplied at a uniform rate. Under such conditions, endothermic and exothermic processes which occur in the sample—by choice the reference is selected to exhibit no changes in the

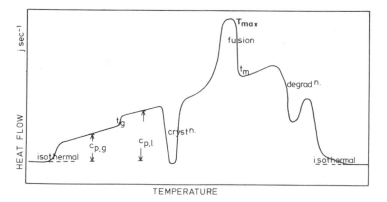

FIG. 6.4. Simulated differential thermal analysis of a quenched glassy polymer which crystallises on heating, exhibiting specific heats, $C_{p,g}$ and $C_{p,l}$, glass transition, T_g, crystallisation, melting and degradation.

temperature range of interest—are reflected by changes in the temperature between the sample, T_S, and reference, T_R, $\Delta T = T_S - T_R$.
Three different forms of DTA have been used:

(1)   classical DTA,
(2)   heat flux differential scanning calorimetry, DSC, and
(3)   power compensation differential scanning calorimetry, see Fig. 6.5.

(1) In a classical DTA unit, the sample and reference materials were inserted into identical environments, heated externally with a common

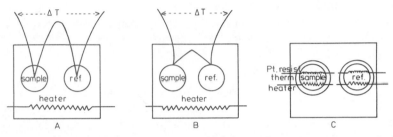

FIG. 6.5.   Differential thermal analysis units. (A) Classical. (B) 'Boersma'. (C) Perkin Elmer DSC.

oven, and the sample and reference temperatures measured by thermocouples mounted in contact with sample and reference. The magnitude of the temperature difference depends on the heat resistance of the system, R, i.e.

$$\Delta T = R\Delta H \qquad (6.8)$$

and

$$\frac{\mathrm{d}}{\mathrm{d}t}(\Delta T) = R\frac{\mathrm{d}H}{\mathrm{d}t}$$

The DTA ordinate (see Fig. 6.4) records ΔT and the abscissa either sample temperature or time. Plotting against programmed temperature, T_p, or reference temperature, T_R, since both increase linearly with time, is equivalent to the latter. The thermal resistance of the systems has several contributions, but the major one arises from sample and reference conductivity which does, of course, vary with temperature but more importantly, varies indeterminately before and after the measured transition. It also varies from sample to sample, with form of the sample and packing density within the sample holder. The proportionality between $\Delta T/dt$ and heat flow, i.e. R, is unknown and ill defined, and peak areas cannot be converted to energies.

(2) Heat flux DSC—To improve the determination of energies Boersma[41] suggested that the temperature sensors should be placed outside, but close to, the sample and reference since this would eliminate the sample variable in the thermal resistance, R, which would then be an instrumental variable which could be measured over the temperature range required. As a result, of course, there is a thermal lag between sample and sensor. Calibration of R against temperature enables ΔT peaks to be calibrated into energies for the transitions.

(3) Power compensation DSC, as developed by Perkin–Elmer Ltd, differs substantially from (1) and (2) in that sample and reference are contained in separate but identical holders with separate heaters and platinum resistance temperature sensors. The powers to the heaters are controlled, one increased and the other decreased by the same amount, according to the instantaneous difference in temperature between sample and reference endeavouring every sixtieth of a second to match T_R and T_S. This is alternated with a test to set the average temperature $(T_R + T_S)/2$ equal to a defined programme temperature, T_p. The ordinate of the DSC trace is the energy difference, determined directly from the electrical power dissipated in the heaters. Although the instrument does require a difference in temperature between sample and reference, in practice ΔT is kept small and T_R, T_S and T_p are effectively the same, and ΔT is minimal. We will see that this is a vital difference when considering exothermic reactions in which heat dissipation becomes controlling. Under such conditions, energy dissipation DSC is less in error than heat flux.

Another obvious advantage, which in practice turns out to have academic rather than practical importance, is that the energy differences are measured directly from differences in electrical power supplied to the heaters, and calibrations are temperature independent. In polymer studies it has been found convenient to use the heat of fusion of an ultra-pure sample of indium as this calibration standard. There is considerable controversy about the actual value of this heat of fusion and until this is resolved it is best to quote the value adopted.

There are differences between the various DTA and DSC instruments commercially available, in their limiting sensitivities, size of sample required, ease of use at low and high temperature, heating and cooling rates available, which make the choice of instrument very much one imposed by the problems studied. In practice, we have observed little difference between the two quantitative DSC systems, and despite the obvious differences outlined above, will treat them as equivalent henceforth.

6.4.1.1. Range of Applications

Differential scanning calorimetry is widely applicable since most physical processes involve either a change in heat content or heat capacity sufficiently large to be detected, if carried out at a sufficiently high rate. Some of the applications are listed in Table 6.4 but this is not intended to be exhaustive but merely indicative of common use. The heat capacity of a polymer can conveniently be measured over a wide temperature range from 150–500 K using liquid nitrogen cooling, but below the glass transition region there is little information on the structure of the glass since the heat

TABLE 6.4
Applications of DSC

General characterisation—'fingerprints'.
Specific heat determination.
Determination of thermodynamic parameters, heat content and entropy.
Glass transition determination, and rates of physical ageing.
Crystallinity, heats of crystallisation, isothermal and non-isothermal rates of crystallisation.
Melting, heats of fusion—crystallite stability.
Decomposition kinetics—thermal and oxidative.
Stability—effects of additives and processing conditions.
Polymerisation kinetics—curing rates and initiator decomposition rates.
Adsorption and desorption—structure of hydrates, etc.
Quality control and effect of additives.
Reaction kinetics.

capacity normally increases progressively with temperature. Low temperature second-order transitions, i.e. β and γ transitions are not observed since there is little or no change in heat capacity at these transitions. Accordingly below the glass transition temperature the heat capacity–temperature curves are quite featureless and contain little information on the structure of the glass.[42] However, the individual values of the heat capacity, $C_p(T)$, vary with the molar composition of the polymer or copolymer system and, for example, there is a small molecular weight dependence since the molar heat capacity per polymer mole,

$$C_p(T) = nC_m(T) + 2C_e(T) \qquad (6.9)$$

where C_p, C_e and C_m are polymer, end-group and monomer units heat capacity, respectively. The corresponding specific heat (per eqn. (6.9)) is then:

$$C_p(T) = C_m(T) + 2\frac{C_e(T)}{(\bar{M}_n)} \tag{6.10}$$

where $\bar{M}_n$ is the number average molecular weight.[15,42]

6.4.1.2. Glass Transition Measurements

Most studies, by DSC, on amorphous polymers focus on glass transition measurements and specific heat changes in the glass transition region where there is an abrupt change (ΔC_p), see Fig. 6.4. The glass transition phenomenon is a non-equilibrium process and is kinetic in nature. Differential scanning calorimetry measurements emphasise these effects and the apparent glass transition temperature (T_g) varies with sample thermal history and rates of cooling and subsequent heating. Accordingly, for a particular cooling rate, used to define a standard glass, the apparent glass transition is heating rate dependent. Richardson and Savill[43] have stressed the importance of defining a standard procedure of defining a glass transition which is independent of measuring conditions, and have suggested that this is best done by measuring specific heat–temperature dependencies well above and below the transition region, since for the glass the heat content at T_g, $H_g(T_g)$ is by definition equal to that of the liquid, $H_l(T_g)$, i.e.

$$H_g(T_g) = H_l(T_g)$$

and

$$H_g^\theta + aT_g + \frac{b}{2}T_g^2 = H_l^\theta + AT_g + \frac{B}{2}T_g^2 \tag{6.11}$$

where a, b, A and B are parameters determined from the corresponding specific heat–temperature dependence, i.e.

$$C_g = a + bT$$

$$C_l = A + BT \tag{6.12}$$

$H_q^\theta - H_l^\theta$ is evaluated by integrating below the heat capacity/temperature curves between two arbitrary temperatures T_1 and T_2, selected well below and above T_g. In this way, T_g is determined from data measured outside the transaction, except for the area under the curve, and so is independent of the pathway from glass to liquid. The T_g measured in this way varies with the rate of heating consistent with its kinetic nature, and the Arrhenius plot of log(rate) against $1/T_g$ gives an estimate of the activation energy of viscous flow, associated with the glass formation.[44]

The glass transition for homopolymer varies linearly with reciprocal number average molecular weight, according to the Flory–Fox relationship,

$$T_g = T_g^0 - A\sqrt{M_n} \qquad (6.13)$$

but various relationships have been discussed recently to correct for end-group effects.

Ageing peaks are also observed in the glass transition region, which appear either on heating the glass through the transition at a rate faster than that at which it was cooled[43] or by leaving the quenched glass at a temperature close to but below the glass transition temperature.[46] In the latter case the size of the ageing peak increases logarithmically with time and appears to be associated with the quenched glass at T_1 moving towards the equilibrium state defined by the enthalpy of liquid–temperature line, see Fig. 6.6. On re-heating, the aged glass follows the enthalpy relationship, AB, but overshoots the glass transition to C when there is sufficient mobility for it to move towards the equilibrium line at D. The enthalpy difference between D and C is equivalent to the ageing peak. Further ageing at T_1 displaces the glass towards the equilibrium line JBD, and produced a

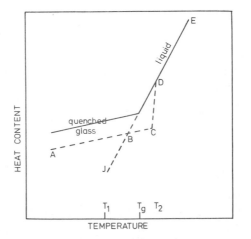

FIG. 6.6.   Temperature dependence of the heat content at the onset of the glass transition.

further superheating of the glass transition and a greater enthalpy change. Accordingly, the aged peaks increase as equilibrium is established and the apparent glass transition is shifted to higher values (see Fig. 6.7), although as defined in Fig. 6.6 the transition temperature is actually lowered.[43] Adopting the procedure outlined by eqns. (6.11) and (6.12)—Richardson and Savill[42,43]—the glass transition is seen to move progressively towards the ageing temperature, T_1.

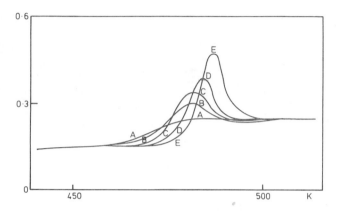

FIG. 6.7. The physical ageing of poly(bis-phenol-A-iso-phthalate copoly-carbonate) at 448 K. AA, quenched; BB, after 1 h; CC, after 2 h; DD, after 4 h; EE, after 16 h. Heat flow mmj s^{-1} against temperature. After C. P. Bosnyak.[16]

Ageing is associated with embrittlement in glassy polymers[46-48] and it is vitally important to establish the temperature range in which it occurs to see if the mechanical properties of a moulded specimen will change with time. It is, of course, unlike other deleterious processes, reversible and the original properties are re-established on heating and quenching for the melt. On ageing the yield stress increases proportional to the size of the ageing peak and there is a linear correlation between the increased energy to deform aged glassy polymer (cf. with quenched) and the size of the ageing peak measured by DSC (see Fig. 6.8), an example in which mechanical properties can be estimated from DSC measurements! Obviously the ageing peak is some measure of the structure of the glass formed.

Above the glass transition temperature the heat capacity of an amorphous polymer liquid increases progressively with temperature with little

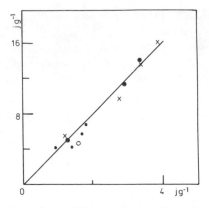

FIG. 6.8.   The variation of the difference in tensile yield energy with the heat of ageing, between the physically aged and quenched polymer. PET, × and ●; PC, ○; PVC, ●. Mechanical energy difference against heat content difference. After A. A. Aref-Azar.[48]

evidence for transitions, unless of course crystallisation, loss of water, or decomposition occurs.

6.4.1.3. Partially Crystalline Polymers

Only poorly crystallised materials exhibit a glass transition temperature, and generally the heat capacity–temperature plots of highly crystalline polymers are quite featureless. There are, of course, well known exceptions to this in that PTFE does exhibit a pronounced transition close to room temperature (c. 8 kJ mol^{-1}). However, the heat capacities are sensitive to structure and are dependent on the degree of crystallinity and vary according to sample preparation.

The mechanical properties, e.g. Young's modulus, yield stress, elongation to break, and strain hardening functions, are sensitive to the extent of crystallisation, such that it is an important material characteristic and DSC and DTA are used widely to measure it. In all calorimetric techniques, crystallinity is assumed to be linearly proportional to the area under the heat of fusion curves (see Fig. 6.4) through a constant considered to be the heat of fusion of the totally crystalline material, $\Delta H_f^\theta(T_m)$ averaged over the melting region of the polymer, i.e.

$$X_c = \Delta H_f(\text{obs})/\Delta H_f^\theta(T_m) \tag{6.14}$$

However, heats of fusion are temperature dependent, and melting occurs over a wide temperature range (5–100 K) and some correction should be made for this, i.e.

$$X_c = \int_{T_1}^{T_m} \left(\frac{dH_{obs}}{dT}\right) dT \Bigg/ \left(\Delta H_{T_m}^0 + \int_{T_m}^{T_m^0} \Delta C_p \, dT\right) \tag{6.15}$$

where $\Delta C_p = C_{p,l} - C_{p,s}$ is the heat capacity difference between liquid and solid.

Relative crystallinities, as determined by eqn. (6.14) however, are consistent, although not absolute, between homopolymers melting in the same temperature range and are easily evaluated if an average value of $\Delta H_f^\theta(T_m)$ can be estimated by calibrating with polymers whose crystallinity has been determined by other means, e.g. density, X-ray diffractometry, etc. However, it should be appreciated that eqn. (6.14) will be in error in comparing polymers which melt in very different temperature ranges.

Copolymers also add a further complication since, if they do not co-crystallise, they restrict the length of chain segments which can, and substantially lower the melting point. On the other hand, if they do co-crystallise, several unit cells will be present and a single valued heat of fusion independent of composition cannot be used.

The observed heat capacities vary with the degree of crystallinity, and amorphous and crystalline regions are considered to contribute proportionately, i.e.

$$C_{p,obs} = C_{p,c}(X_c) + C_{p,l}(1 - X_c) \tag{6.16}$$

in which X_c is a weight fraction degree of crystallinity. Measurements above the m.pt. enable $C_{p,l}$ to be determined as a function of temperature. Linear extrapolation to the temperature range of interest enables $C_{p,l}$ to be evaluated. $C_{p,c}$ is evaluated by measuring $C_{p,obs}$ for polymers of different, but known, degrees of crystallinity.

As the temperature increases the polymer, if it does not decompose, will melt and the endotherm, unlike low molecular weight materials, occurs over a wide temperature and varies according to crystallisation temperature and molecular weight of the polymer. The melting point of a polymer is not unique and has little equilibrium thermodynamic significance. Indeed, it is possible to re-crystallise a polymer above its apparent melting point and somewhat special handling of the melting point data is required if meaningful thermodynamic parameters of melting have to be determined. It is common practice with low molecular weight material to use the maximum rate of melting, T_{max}, see Fig. 6.4, to define a melting point.

However, this is wrong for a polymer, since in a partially crystalline chain compound there is a range of lamellar crystals of varying thicknesses, ξ, and different degrees of perfection. The melting point of a thin crystal depends on its thickness and lateral surface free energy, σ_e,[49] i.e.

$$T_m = T_m^0[1 - 2\sigma_e/\Delta h \xi] \qquad (6.17)$$

in which Δh is the heat of fusion per repeat unit, there being ξ repeat units in the thickness, and T_m^0 refers to the infinitely extended chain crystal's melting point.

Clearly the melting range contains information of the range of lamellar thicknesses within the polymer sample and does not necessarily reflect thermal lag effects between thermal sensors and melt temperatures. Choosing T_{max} as the melting point is equivalent to taking the melting point of the commonest lamellar crystal thicknesses in the distribution. We have adopted Flory's procedure[50] of choosing the last trace of crystallinity as defining the melting point of the more stable crystallites in the sample and correcting for thermal lag and sample size effects by varying the weight and heating rate. Under these conditions the melting–temperature curves are considered to contain some measure of the lamellae distribution and indeed to be a measure of the quality of the crystallinity within the spectrum. Melting and re-crystallising the sample at higher temperature increases the critical size nuclei, as determined by classical nucleation theory,[51] and so thicker lamellae are produced. The melting range increases with crystallisation temperature. Hoffmann and Weeks[51] have formalised this, and derived the dependence of observed melting point, T_m, on crystallisation temperature, i.e.

$$T_m = T_m^0(1 - 2/\beta) + T_c/2\beta \qquad (6.18)$$

in which $\beta = 1$ if the crystallisation and subsequent melting do not involve annealing effects which thicken the crystals. Plots of T_m against T_c are linear, with slopes of 0.5, and can be extrapolated to cut the equilibrium line $T_m = T_c$ at T_m^0 (the equilibrium melting point of the polymer); this melting point is independent of crystallisation conditions. This procedure, however, is only valid if the molecular weight is 10^5 and over.[52]

Afifi–Effat[53] observed that under non-equilibrium conditions, i.e. $\beta \neq 1$, the observed slopes of linear plots of T_m against T_c determined by DSC were simple multiples, since $\beta = \xi_e\sigma/\xi\sigma_e$ where subscript refers to equilibrium conditions, so that the impression was gained that in fact lamellae thickened in multiples of the original lamellar thickness, ξ.

The melting points of polymers of low molecular weights are determined by molecular chain length[54] if it is below a certain value; 200–400 —CH_2— repeat units in the case of polyethylene. Under these conditions, the lamellar thickness is limited by the number of repeat units, n; then the free energy of formation of the lamellae crystal is

$$\Delta G^* = n\Delta g_f + \Delta g_e - 2RT \ln (n) \qquad (6.19)$$

in which Δg_f, Δg_e refer to free energy difference of monomer and end-group units in the solid and liquid, and the last term $RT \ln (n)$ the entropy of mixing end-groups ($x = 2$,[54] or $x = 1$[55]). This equation is limited to monodisperse oligomers and can only be applied to polydisperse systems by including a term $RT \ln (I)$ for the polydispersity.[56] Therefore, if

$$\Delta g_f = \Delta h - T\Delta S$$

$$\Delta g_f = 0 \qquad \text{at } T_m^0$$

and

$$\Delta G^* = 0 \qquad \text{at } T_m$$

then

$$T_m = T_m^0 \left[1 - \frac{2RT_m^0 \ln (n)}{\Delta hn} \right] - 2\sigma_e / n\Delta h \qquad (6.17a)$$

which is somewhat similar to eqn (6.17) except that it includes an addition term for the entropy of mixing of terminal units. Both equations will not of course derive the same value for σ_e. Plots of T_m against $\ln (n)/n$ are linear, for polyethylene, n-alkanes, poly(ethylene oxide)[54] and other systems, and appear to be a valid description of the dependence of T_m on chain length (n). The slope and intercepts further enable T_m^0 and $\Delta H^\theta(T_m^0)$ to be determined, which compares favourably with other estimates.

The lamellar size distribution can be estimated by other means, low angle X-ray scattering, longitudinal acoustic mode in the Raman spectrum and nitration of the polymer which etches out the crystalline regions and GPC analysis of the products. Equation (6.17a) has been used recently[57] to interpret the distribution in lamellae from the melting temperature–heat flow curves obtained from DSC and compared with expected GLC distribution obtained on nitration, see Fig. 6.9. Reasonably good agreement was achieved between the two methods, considering that no allowances were made for crystal imperfections within the partially crystalline polymer which would lower the melting points. Clearly this approach requires extension to other systems, but it shows promise that

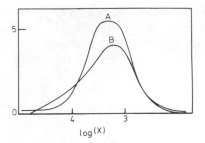

Fig. 6.9. Stem length distributions, determined by: (A) nitration and GPC, and (B) analysis of the melting curves. Relative weight against stem length degree of polymerisation, X_n.

thermal analysis can give information about the quality and nature of the crystallites present in polymer systems.

Differential scanning calorimetry is widely used to measure the development of crystallinity isothermally with time, but it is severely limited by the sensitivity of the ΔT detectors, such that crystallisation rate studies in the case of polyethylene are restricted to a temperature range of 4–7 K corresponding to half lives of 2–200 min. Obviously this varies from polymer to polymer. No such restrictions are usually observed with dilatometry and so isothermal crystallisations measured by the two techniques will be considered later.

In practice crystallisation is seldom encountered under isothermal conditions but usually occurs during quenching or slow cooling within the mould. DSC and DTA, with the facility to cool the melt at various rates and follow the development with temperature (or time), are particularly meaningful, since they enable the effectiveness of various added nucleating agents to be assessed,[58] in the temperature range in which crystallisation proceeds in practice, to determine crystallisation temperatures and effect on stability of the crystallites produced. Non-isothermal crystallisation kinetics have not been particularly successful in describing such crystallisations[59] but finite element analysis using isothermal rate constants corresponding to each temperature[60] indicates that heat losses from the sample are rate determining, unless cooling rate is low. Accordingly, crystallinity most likely develops from the surface of the sample inwards and surface/volume ratio may be more important than nucleation or growth rates.

Nucleation phenomena have been observed by DSC during the crystallisation of polyethylene blends[61] and block copolymers[62] which seemed to exhibit anomalous crystallisation characteristics. By dispersing poly-ethylene as fine droplets in amorphous polystyrene[61] the size of the droplets increased with PE content. It was observed that the isothermal crystallinity of PE developed rapidly (within seconds) and was limited by the temperature. The apparent crystallinity increased progressively on cooling but not with time isothermally. This was considered to be due to the number of nuclei developing at each temperature being less than the number of droplets present. Crystallisation within a droplet occurs rapidly if an active nucleus is present for that crystallisation temperature. Knowing the average size of the droplets and the fraction of them crystalline, enabled the number of heterogeneous nuclei to be estimated at each temperature. Only at high supercooling, 40–60 K, was homogeneous nucleation of PE observed. Nucleation of bulk crystallisation, at relatively low degrees of supercooling, by analogy must be heterogeneous.

6.4.1.4. Reaction Rate Studies

Thermal analysis is widely used to study thermal and oxidative de-gradation, rates of curing, and polymerisation and initiator decom-position, but in all these studies DTA and DSC are limited to compara-tively rapid rates since invariably micro-samples are used, the heat of reactions are finite, both methods are measuring a rate of change and have a limiting sensitivity. Since micro-samples are involved, encapsulated in aluminium pans, there is a further limitation on purity and eliminating last traces of moisture, oxygen, etc., which severely limit the number of kinetic rate studies which can meaningfully be made.

The cure characteristics of epoxy resins have been studied by Barton[63] in open aluminium pans and in a stream of nitrogen, both isothermally and at various heating rates. Samples were held at constant temperature until cure was apparently complete, when an experimental base-line could be determined for extrapolation to zero time. Rates were measured from deflection for the base-line as a function of time.

The integrated extent of reaction could not be fitted to a general rate expression:

$$\mathrm{d}d/\mathrm{d}t = k(1 - \alpha)^n \qquad (6.20)$$

in which d is the degree of conversion, k the observed rate constant and α the extent of conversion. A rate equation, derived by Sounour and Kamal,[64] could be fitted, i.e.

$$\mathrm{d}d/\mathrm{d}t = (k_1 + k_2\alpha)(1 - \alpha)^2 \qquad (6.20a)$$

to above 50 % conversion.

No kinetic model could be obtained which fitted the data better. However, the isothermal conversion–log(time) plots could be superimposed along the log(time) axis, and a master curve could be determined.

The thermal decomposition of polyacrylonitrile has also been separately studied by DSC[24b] and DTA[25] and highlights an important problem area for thermal analysis, namely highly exothermic reactions. Kennedy and Fontana[65] and later Thompson[66] reported that an exothermic decomposition reaction evolving volatiles occurred in the temperature range 620–700 K. The decomposition reaction was exothermic to 5–6 kcal mol^{-1}. The isothermal reactions appeared to be autocatalytic with pronounced induction periods.[24] A DTA study,[26] however, indicated that the heat evolved was accelerating the reaction and large temperature differences, up to 15 K, were developing between sample and reference. The progress of the reaction was controlled by the rate of loss of heat to the surroundings and with a good thermal insulator, such as polyacrylonitrile, the isothermal reaction extent–time curves were undoubtedly determined at the higher temperatures by the progressively increasing temperature of the specimen. Accordingly, both DSC and DTA are limited in studying highly exothermic reaction to low rate and small samples dispersed between good insulators to disperse the heat. Film thickness and particle size effects, on the kinetic rate constants and progress of the reaction, are evident.

6.4.2. Adiabatic Calorimetry

The adiabatic calorimetry occupies a unique position in classical thermodynamic studies of polymers in the measurement of accurate values for specific heats, heat contents and absolute entropies,[67] but appears to have lost its popularity to more rapid, and less accurate, methods of determining thermodynamic parameters. Adiabatic calorimetry has established much of our understanding of glass formation, and crystallisation–melting transition, and still remains the most accurate of thermal analytical techniques.

Dynamic adiabatic calorimeters are also available commercially for quantitative study of polymeric thermal characteristics, in which either the temperature of an isolated cell containing the sample is raised linearly with time and the electrical power supplied monitored, or else the power supplied is constant and the temperature monitored with time. Both are equivalent, and knowing the thermal capacity of the sample container, the specific heat or heats of transition can be determined directly.

One particular advantage of calorimeters, in general, is that they have been designed with sufficient capacity for dynamic experiments, such as fracturing or yielding, to be carried out on the specimens. When a polymer specimen yields under tension by fracturing a cold-drawing heat is generated and there has been considerable speculation about the source of this heat and its importance during the deformation and production of a neck. Calorimetric studies of the necking of polymers have the distinct advantage that the total heat generated can be directly related to mechanical energy. Using this procedure, Müller[68] reported that the ratio of mechanical work to heat generated for polycarbonate was 8:5 and Andrianova et al.[69] for poly(ethylene terephthalate) a ratio of 1. Polypropylene has been observed to be anomalous with a high ratio of 10:1, presumably due to integral energy changes occurring in the crystalline polymer. A similar study has been carried out recently on a wide range of specimens using an infrared camera—Aga Thermovision System 680—to determine the temperature profile which develops directly in a dynamic neck.[70] Direct heat measurements could not be measured by this technique and corrections had to be made for heat losses to the surroundings. The i.r. detector was an indium antimonide photovoltmeter and the infrared radiation was displayed on a video display unit which gave a real time thermal picture of the temperature profile within the neck—see Fig. 6.10. Temperature measurements were calibrated against a black body radiator with known surface temperature, and to ensure that the necking plastic was opaque to i.r. radiation studies were made on black pigmented material. Measurements were made of the maximum temperature rise within the neck, against velocity of neck propagation and from a consideration of energy balance a relationship between reciprocal temperature rise and cross-head velocity, V_c, i.e.

$$\frac{1}{\Delta T} = \frac{kA}{Lgd}\frac{1}{V_c} + \frac{ab\rho C_p}{(D-1)Lgd} \tag{6.21}$$

where k is heat loss from unit surface area, A, a, b and L are dimensions of the specimen and D the draw ratio. Plots of $1/\Delta T$ against $1/V_c$ were linear and from these the ratio of mechanical to heat energy d was determined, Fig. 6.11. In most cases studied d was above 1·0 but within the experimental uncertainty, indicating that the mechanical energy accounted for the temperature rise in the necking region. Extrapolation to infinite velocity is analogous to ideal adiabatic conditions, with no heat losses to the surroundings. The thermovision has the advantage of studying a mechanical/dynamic process within a polymer remote from it and not interfering with its progress.

Finite element analysis[71] has been applied to the generation of heat

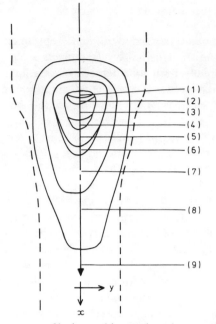

FIG. 6.10. Temperature profile in necking polycarbonate during extension as measured by i.r. camera. After J. W. Maher *et al.*[70] (1) 39·3 °C, (2) 37·3 °C, (3) 35·3 °C, (4) 33·5 °C, (5) 31·5 °C, (6) 29·9 °C, (7) 27·7 °C, (8) 25·6 °C, (9) 23·0 °C.

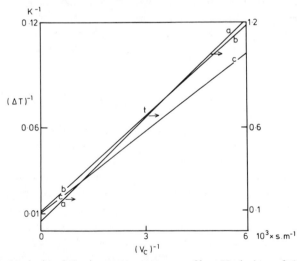

FIG. 6.11. Analysis of the i.r. temperature profiles. Variation of the maximum temperature rise with cross-head velocity, V_c. a, PVC; b, PE; c, PP. After J. W. Maher *et al.*[70]

assuming complete conversion of mechanical energy during necking and correcting for thermal losses from the surface of the polymer. A similar thermal profile is computed to that observed by the thermovision. The thermovision is an interesting extension of the thermal measuring devices which are available to the polymer chemist. With improved detection limits this instrument could become increasingly important in thermal analysis, especially in examining temperature profiles in isothermal samples undergoing exothermic or endothermic changes. Further application of this infrared detector to polymer systems, especially in measuring thermal conductivity and heat diffusion will occur.

6.5. DIMENSIONAL CHANGES

6.5.1. Dilatometry
Thermodilatometric analysis is the continuous monitoring of the length, area or volume of a specimen as a function of temperature or time isothermally. The technique is a classical physical one and has been used widely in polymer science to study many processes, e.g. polymerisation kinetics, glass transition temperature, melting and crystallisation kinetics. Many of these original studies were made on the simplest of apparatus on polymer samples stored over mercury in glass dilatometers, and heated externally with a thermostatted liquid bath whose temperature could be held accurately at a fixed temperature or heated slowly at a constant rate. Most, however, would exclude this simple technique as not part of thermal analysis and would restrict the term to those instruments which automatically record dimensions and have programmed temperature control. This in itself still does not exclude the glass dilatometer. Most modern thermodilatometers monitor the linear expansion of a solid or glass sample through a mechanical push rod which lightly rests on the free end of a sample. The sample is heated externally with an oven. The push rod's weight is compensated by a support system which balances the weight and only a nominal pressure of less than 1 g is exerted on the sample. For most accurate and absolute measurements a differential system is adopted in which the expansion of the sample is compared to that of a standard. Liquid samples exhibit a problem, and several commercial instruments can only be used with solid/glass samples. However, it is possible to modify the Linseis liquid sample holder to most instruments. This involves encasing the sample in a light invar metal cylinder into which the push rod makes a tight fit.

In most commercial dilatometers the measuring device is a linear variable differential transformer which responds to the end of the push rod. Other methods have been adopted of measuring the deflection, e.g. strain gauges, and mechanical levers. These have been reported elsewhere.[6] These instruments are particularly sensitive to linear dimensional changes and so are useful in detecting small changes associated with the onset of transitions within the polymer. Specimens of 3–6 mm can conveniently be handled on the Stanton-Redcroft, TMA 691, and changes in length 10^4 times less can readily be detected. This is of the order of the dimensional changes associated with physical ageing (see above) and this instrument has been used to follow the contraction in volume (length) associated with the densification of the glass towards the equilibrium free volume.

As with the thermal analysis study of ageing the dimensional changes decreased logarithmically with time. However, this indicates a problem which is inherent in all recording systems, but not observed in the glass–visual reading dilatometer, i.e. the eventual lack of sensitivity and long-termed stability of the LVDT, which must become limiting in a logarithmic dependence on time. It is inconvenient to use such recording dilatometers over several days, weeks or months, and so inherently the high sensitivity and great potential of these techniques to obtain accurate measurements, with very slow rates of heating or isothermally, is lost.

Dilatometry has been widely used to follow the isothermal rate of crystallisation of polymers (see Fig. 6.12) particularly when they are very slow to crystallise and so there is very little rate of heat flow. It can be used when DSC or DTA would not detect any crystallisation. These slow crystallisations can be followed by slow rates of heating, e.g. 1 K day^{-1} and 1 K month^{-1}, to determine the subsequent melting characteristics of the polymer.[73] By heating at 1 K day^{-1} or 1 K month^{-1} it was assumed that the polymer could be melted and recrystallised at each temperature such that the final trace of crystallinity would persist at the equilibrium melting point. In practice even with these very slow rates of heating, equilibrium is not achieved and the melting points observed are considerably lower than the equilibrium value, 139·5 °C in the case of PE compared with a T_m^0 value of 145 °C[74] and 69·5 °C in the case of poly(ethylene oxide) compared with a T_m^0 value of 76 °C.

A partially crystalline polymer is considered to consist of two phases, a crystalline one with specific volume V_C, and amorphous one of specific volume V_A. The observed specific volume varies according to the degree of crystallinity, X_C, i.e.

$$V_{obs} = (1 - X_C)V_A + X_C V_C \qquad (6.22)$$

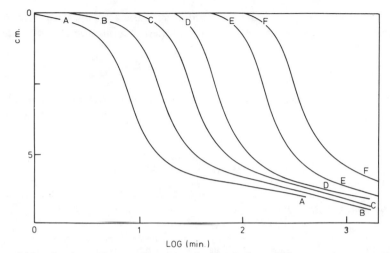

FIG. 6.12. Isothermal crystallisation of polyethylene. Dilatometric contraction against log time. 200 mg of Marlex-50 fraction. A, 399·0 K; B, 400·2 K; C, 401·3 K; D, 401·9 K; E, 402·4 K; F, 402·9 K.

V_A values are determined by linear extrapolation of the melt to the temperature of interest, and V_C from crystallographic unit cell dimensions. Equation (6.22) defines a degree of crystallinity, i.e.

$$X_C = (V_{obs} - V_A)/(V_C - V_A)$$

and is made use of in dilatometry to measure the degree of crystallinity as a function of time.

The progress in crystallinity can be followed by the linear dimensional change, since:

$$(V_{obs} - V_A)/(V_C - V_A) = (h_t - h_0)/(h_c - h_0) \qquad (6.23)$$

where $h_c - h_0$ is the contraction calculated for a totally crystalline sample, and h_0 and h_t are the observed readings (initially and at time t). The polymer produced in the sample is, however, not totally crystalline, and so absolute crystallinities are usually replaced by relative crystallinities at time, t, i.e.

$$X_C' = (h_t - h_0)/(h_\infty - h_0) \qquad (6.24)$$

where $h_c > h_\infty$ and h_∞ is the limiting value determined experimentally.

Accordingly, the Avrami equation relating relative crystallinity to time, t, is:

$$-\ln(1 - X_C) = Zt^n \qquad (6.25)$$

in which Z is a composite rate constant and n a mechanistic parameter but restricted to integer values 1 to 4, is extended to dilatometry, such that

$$-\ln\left\{\frac{(h_\infty - h_t)}{(h_\infty - h_0)}\right\} = Zt^n \qquad (6.26)$$

and an average value of the Avrami exponent is then determined from plots of $\log(-\ln\{(h_\infty - h_t)/(h_\infty - h_0)\})$ against $\log t$, and $\log(Z)$ from the

TABLE 6.5
Avrami Parameters for Crystallisation

(a) *Predicted values*

Mechanism	Rate (n)	Parameters (Z)	Restrictions on geometry
Sporadic spheres	4·0	$\frac{2}{3}\pi\dot{n}\dot{g}^3$	Three-dimensional growth
Predetermined spheres	3·0	$\frac{4}{3}\pi N\dot{g}^3$	Three-dimensional growth
Sporadic discs	3·0	$\frac{\pi}{3}d\dot{n}\dot{g}^3$	Two-dimensional growth (restricted to a surface)
Predetermined discs	2·0	$\pi d N\dot{g}^3$	Two-dimensional growth (restricted to a surface)
Sporadic rods	2·0	$\frac{\pi}{4}D^2\dot{n}\dot{g}$	One-dimensional growth (restricted to a capillary
Predetermined rods	1·0	$\frac{\pi}{2}D^2 N\dot{g}$	One-dimensional growth (restricted to a capillary)

$\dot{n}$, $\dot{g}$ are nucleation and radial growth rates, N is the nucleation density, d is the thickness of surface and D is the radius of the capiliary.

(b) *Observed in practice*[78]

Polymer	Range of n value
Poly(ethylene oxide)	2–4
Polymethylene	1·8–2·6
Polyethylene	2·4–4·0
Polypropylene	2·8–4·1
Polystyrene	2·0–4·0
Poly(decamethylene terephthalate)	2·7–4·0
Poly(ethylene terephthalate)	2·0–4·0

intercept at $t = 1$, or alternatively instantaneous values can be obtained from

$$n = -t^{(\mathrm{d}h_t/\mathrm{d}t)}/(h_\infty - h_t)\left[\ln\frac{(h_\infty - h_t)}{(h_\infty - h_0)}\right] \qquad (6.27)$$

and Z is calculated using n value at 50 % conversion and the half life $t_{1/2}$ since:

$$ZA = \ln(2)/(t_{1/2})^n$$

As with all kinetic schemes the validity of the analysis rests on the reproducibility of the kinetic parameters, i.e. Z and n values, and their actual values since they are diagnostic of the mechanism, see Table 6.5. In particular, most polymers crystallise as spherulites and so n is restricted in value to 3 and 4. In general, fractional values are observed (see Table 6.5) which have little or no significance. This, however, is not a feature of dilatometry. Differential scanning calorimetry[75,76] and other techniques[77] give very similar fractional n values. In particular, comparison of DSC and dilatometry on the same polymer sample showed that little or no difference exists between the two techniques, although in fact the rate parameters were meaningless in terms of the crystallisation models adopted by Avrami, see Table 6.6.

Dilatometry is an accurate method of studying crystallisation kinetics since the individual crystallinities can be estimated to better than 0·1 %. Despite this, the uncertainties in the rate expression to describe the

TABLE 6.6
Comparison of Dilatometry with DSC. Polyethylene Crystallisation[75,76]

Temperature	n value	$t_{1/2}$ (min)	Z (min^{-n})
(a) Dilatometry			
400·7	2·8	21	62×10^{-6}
401·6	2·9	40	7·2
402·2	2·8	70	2·9
403·5	2·8	114	1·2
(b) DSC-1B			
400·7	2·7	20	66×10^{-6}
401·7	2·8	36	7·2
402·7	2·6	68	3·2

crystallisation data restrict the application of this procedure to determine meaningful heats of fusion and lateral surface free energies. The composite rate constant, Z, exhibits a dependence on the supercooling, i.e. $(T_m^0 - T_c)$, which is quite complex since both $\dot{n}$ and $\dot{g}$ have their own dependencies, but the fractional value of n parameter makes the evaluation of Z uncertain. That is,

$$g(T) = g_0 \exp(+\Delta G_d/RT) \exp(-\Delta G^*/RT) \qquad (6.28)$$

in which g_0 is a frequency factor and shape constant, ΔG_d is the free energy of bulk diffusion of the segments across the liquid/solid boundary and ΔG^* is the free energy of formation of the critical size nuclei, and is usually rate determining. Also,

$$\Delta G^* = 4\sigma_u \sigma_e T_m^0/\Delta h(T_m^0 - T_c) \qquad (6.29)$$

where σ_u is the surface free energy of the non-lateral faces. $\ln(Z)$ usually displays a linear dependence on $(\Delta T)^{-1}$, which confirms the analysis in terms of nucleation dependence of growth, but leads to uncertain values for σ_u and σ_e.

Ashman et al.[89] have also placed a dilatometer inside a Calvet calorimeter and followed the same crystallisation dilatometrically and calorimetrically. They observed that low molecular weight poly(ethylene oxide) crystallises at the same rate as measured by calorimetry and dilatometry, but high molecular weight fractions lagged behind at higher degrees of crystallinity. They interpreted this in terms of voids forming in the bulk between spherulites and the melt, not diffusion into these.

An attempt was made to modify[80] the Avrami equation by including a visco-elastic term with a logarithmic time dependence to allow for this and also account for a slow secondary process which appears to have a logarithmic time dependence. Accordingly,

$$X_C = X_{p,\infty}(1 - \exp - Z_p t^n) + X_s \int_0^t [(1 - \exp - Z\theta^n)/(t + \theta - 1)]\,d\theta$$
$$(6.25a)$$

in which p and s refer to primary and secondary processes, respectively.

However, correcting for the secondary process did not substantially improve the degree of fit of the crystallisation–time dependence and again the best overall description involved fractional n values—similar in fact to those determined by a fit of the simpler Avrami equation to the initial portion of the crystallisation–time dependence.

Crystallisation kinetics is waiting for a better kinetic description of the overall process. When this kinetic expression is developed many analytical techniques are available to test this to its limit. Not least of these techniques will be dilatometry.

6.5.2. Thermomechanical Analysis

If an increasing load is applied to a polymer sample, in tension or under compression at constant temperature, it will deform elastically (initially) and then yield when permanent deformation occurs, see Fig. 6.13. The

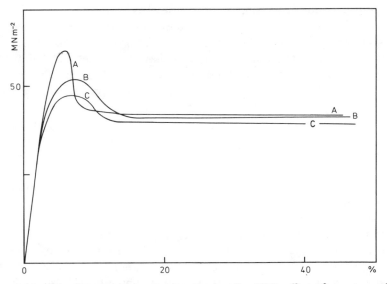

FIG. 6.13.   Tensile engineering load–extension for PVC; effect of structure. AA, amorphous—aged; BB, 5% crystalline; CC, amorphous—quenched.

load–deformation curve is an important parameter of all materials since it defines Young's modulus, yield stress, uniform drawing or necking— brittle, weak or ductile behaviour. These properties are temperature dependent, and brittle materials become more or less ductile as the glass transition is approached, and using an environmental chamber it is possible to determine stress–strain curves as a function of temperature. Thermo-mechanical analysis attempts to determine the effect of temperature on mechanical properties by continuously varying the temperature and to do this the sample is subjected to a constant strain. It is an attempt to measure single point mechanical properties as a function of temperature. The full load–extension (or contraction) curve can only be determined from a series of such determinations up to the yield point. Above this, deformation is non-elastic and time-dependent. Useful information is sought from the temperature dependence of the way materials react to constant loads.

Compression studies are determined using dilatometers with accurately balanced push rods which exert no positive pressure on the sample. Known weights (up to 100 g) are added to the end of the push rod and the deformation of the specimen determined as a function of temperature and applied load. Push rods, 1 mm^2, can be used with stress of several MN/m^2 locally applied to the specimens and continuous change in loading can be examined isothermally. However, it should be appreciated that the load on the specimen is highly localised and not uniform overall. Push rods with diameters similar to that of the specimens have to be used to simplify the distribution of pressure throughout the specimens—this limits the applied stress to 0·1 MN/m^2. Small contact areas of the probes can be used qualitatively to detect 'softening' of the polymer by measurement of the temperature of penetration, which will vary considerably with applied load. However, as with dilatometry, TMA is used to evaluate coefficients of expansion, and these are sensitive to transitions within the polymers, α, β and γ transitions and also melting points. This has been used by Haldon et al.[81] to detect transitions in glassy polymers. The Vicat softening temperature, heat distortion temperature, or sample deflection temperature are readily measured.

Melting of crystalline polymers is not sharp and length changes followed by penetration occurs over a wide temperature range. These generally occur well below other measurement of melting points and must really be considered measurement of softening points. These heat distortion temperatures do have technical significance.

Studies can be made on thin strips of polymer film or on fibres by clamping the two ends—one to a firm support and the other to the push rod. Three point bending tests can also be made on rigid samples held on knife edges and bend, centrally between the edges, by the push rod. However, because of the restriction on sample dimensions—it is not possible to use specimens which are dumbbell shaped—there must be considerable specimen end effects, and so these studies must be considered qualitative.

Thermomechanical analysis can, however, be made quantitative if the mechanical stage is substantially improved and strain gauges are mounted in the test-rigs. This, however, limits the study either to isothermal studies[82] or low rates of heating with constant applied stress because of the greater thermal mass of the substantial specimens and mass of the stage. Recently Gilmour *et al.*[83] have measured the thermoelastic effect in compression on several glassy polymers and obtained quantitative results ($\pm 2\%$) by paying attention to specimen geometry and end effects. The polymers obeyed the classical Thompson equation:

$$\Delta T = -\alpha T \Delta \sigma / \rho C_p \qquad (6.30)$$

in which ΔT is the thermoelastic temperature change within a specimen to which a uniaxial tensile stress, $\Delta \sigma$, is applied at temperature T. The density is ρ, C_p is specific heat, and α the coefficient of linear expansion.

The instrument included a furnace whose temperature could be controlled isothermally or heated or cooled linearly in the range 210–370 K, which fitted between the runner bars of a Monsanto tensometer, under a dry inert atmosphere. The samples were in the form of cylinders with an encapsulated thermocouple. Instantaneous application of a compressive load (up to $40\,\mathrm{MN\,m^{-2}}$) produced a rise in temperature (up to 0·04 to 0·8 K) which increased linearly with load. These thermoelastic measurements enabled $\alpha V / C_p$ to be evaluated for the glassy polymer, and from the bulk modulus (β_T) the thermodynamic Grüneisen constant (γ_T) can be evaluated, since

$$\gamma_T = 3 \beta_T \alpha V / C_v \qquad (6.31)$$

When a solid receives a pressure point[84] it will generate a pressure response; this is related to the Grüneisen constant which is defined as

$$\gamma = V \left(\frac{\mathrm{d}P}{\mathrm{d}E} \right)_u \qquad (6.32)$$

This constant is a most important material property, and using parameters from the thermoelastic response, Gilmour *et al.*[84] were able to determine the constant over a wider temperature range than previously.

6.6. DYNAMIC THERMOMECHANICAL ANALYSES

These techniques measure mechanical properties on samples under an oscillatory load and are extremely useful technically in considering the overall performance of polymers, since the mechanical properties are measured as a function of temperature and frequency. These properties are also very sensitive to molecular structure, and changes with temperature have been useful in determining glass transition region, the presence of crystallinity, phase separation and cross-linking. Polymer blends, block copolymers and copolymerisation have also been analysed by dynamic thermomechanical analysis.[85]

There are many types of instruments available, commercial and research, which differ in the type of measurement made and the frequency of the loads applied. Some measure of Young's modulus, others shear and even the bulk modulus. Each technique has its own limitations and restrictions in use, and also range of applications, and is best considered separately under the different headings, but only a few can be considered.

In general, polymers are viscoelastic and as such, in deforming, energy is stored elastically, as potential energy, and some is dissipated as heat. The energy lost as heat manifests itself as mechanical damping. For each test, an elastic modulus and mechanical damping are measured as a function of temperature and oscillatory load frequency. To exploit the full potential of the techniques, measurements should be made over as wide a range of temperatures and frequencies as possible. Within the limitations of the technique, however, it is generally better to choose low frequencies, e.g. 1 Hz, rather than high frequencies, 10^4–10^6 Hz. Secondary transitions are more readily detected at low frequencies.

6.6.1. Torsional Pendulum

This measures the shear modulus of a strip specimen, clamped rigidly at one end, and the other attached to an inertial arm which is free to oscillate. Oscillations twist and untwist the specimen with an oscillation time period, t_p, i.e. the time to complete one oscillation. Damping converts the mechanical energy to heat and the amplitude of the oscillations decrease. The shear modulus, G, is derived from t_p, since

$$G' = C/t_p^2 \qquad (6.33)$$

where C includes sample dimensions, polar moment of inertia of the system, and a shape factor.

The damping, expressed as the logarithmic decrement, is measured from the decreasing amplitude with time. It is the ratio of the amplitudes of two consecutive oscillations, i.e. A_1, A_2, A_3, etc.

$$\Delta = \ln \frac{A_1}{A_2} = \ln \frac{A_2}{A_3} = \cdots \text{etc.} \qquad (6.34)$$

For very high damping, i.e. $\Delta > 1$, G' depends on the damping and the above equation has to be modified to

$$G' = \frac{C'}{t_p^2} [4\pi^2 + \Delta^2] \tag{6.35}$$

The sample may be under a net tension, and the above equations are only valid for no net tension. The period of oscillation is corrected by extrapolating $1/t_p^2$ against added tensile loads to zero loads.

The torsional pendulum is restricted to a narrow range of frequencies normally 0·01 to 10 Hz, but is applicable to the entire range of modulus, 10^4–10^{10} N m^{-2} and a range of damping from 0·01 to 5, encountered in polymers.

6.6.2. Torsional Braid Analyser[86]

This has been applied mainly to polymerising systems, and polymers. The sample in the above torsional pendulum is replaced by a glass or carbon fibre braid. The braid is impregnated with the sample in solution or a polymerising resin. The torsional effects arise from the braid and the sample. As the polymerisation develops so there will be an increase in modulus and a decrease in damping. Glass transition, crystallisation and melting, and subsequent decomposition of the polymers coated on the braid will be accompanied by a decrease in rigidity and increase in damping[87] but it is a relative measurement only.

6.6.3. Vibrating Reed[88]

This is widely used to measure damping and Young's modulus. The polymer sample is a thin reed clamped firmly at one end and made to vibrate transversely, with a variable frequency oscillator. Frequencies from 10 to 10^3 Hz can be used. As the frequency is changed, the reed will oscillate at a natural resonance, and the amplitude of the free end reaches a maximum. The change in amplitude with frequency about this natural resonance is measured, and the modulus (E) calculated, since

$$E = C\rho L^4 f^2 / D^2 \tag{6.36}$$

where ρ is the density, L the length, D the thickness and f the resonance frequency. The damping is obtained from the half-width of the frequency–amplitude curves. It is more often measured from the amplitude of the clamped and free ends of the reed at the resonance frequency. Since this is a resonating reed, both temperature and frequency have to be varied—for a single specimen it is not possible to study the effect of frequency at constant temperature, or temperature at constant frequency.

A rotating rod[89] can also be used. It is clamped at one end and rotated along its length. A transverse force is applied to its free end which bends it, but because of the rotation this is at an angle to applied force. Mechanical damping is the tangent of this angle. Young's modulus is measured from the applied force and deflection.

6.6.4. Rheovibron, Viscoelastometer and Dynamic-mechanical Thermal Analyser

There are many other types of non-resonating measurements in which the amplitudes of the stress and strain are measured separately and the phase angle difference between the stress and the strain determined. A mechanical vibrator delivers a sinusoidal tensile strain to one end of a clamped specimen; this is relayed through the specimen and the phase and amplitude of the displacement measured at the other end of the specimen. Two separate transducer systems are used to measure amplitude of the stress and the strain, from which the absolute values of the modulus and $\tan \delta$ are computed. The symbol δ represents the phase angle between the strain and the stress, and is a measure of the ratio of the imaginary parts, E'' to the real, E', parts of the complex modulus, E^*,

$$|E^*| = |\sigma|/|E| \qquad \text{i.e. stress/strain}$$

and

$$E' = |E^*| \cos \delta, \qquad E'' = |E^*| \sin \delta \qquad \text{and} \qquad E''/E' = \tan \delta$$

E'' is the damping or loss modulus.

Commercial instruments operate at different ranges of frequencies and with different sensitivities. The Rheovibron Viscoelastometer (model DD-II) operates from 3·5 to 110 Hz and the PLDMTA 3·3 × 10^{-2} to 90 Hz. Most instruments cover a limited range, but instruments in general can be matched to cover 10^{-3} to 10^4 Hz, but not one single unit.

Experiments are carried out at constant frequency over a range of temperatures where increases in $\tan \delta$ are associated with transitions in the polymer, see Fig. 6.14. In addition to the main glass transition due to motions of large segments of the polymer chain, many polymers show secondary transitions, which are normally attributed, in amorphous polymers, to motion of side groups. Although these transitions are not detectable by DTA and DSC they are readily detected by damping measurements.

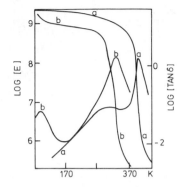

FIG. 6.14.   Dynamic mechanical spectrum of poly(methyl methacrylate); 11 Hz. a, poly(methyl methacrylate); b, poly(n-butyl methacrylate). After P. J. Mills, private communication.

Poly n-alkyl methacrylates have transitions[90] which vary somewhat with ester structure. The α transition, or glass transition, decreases progressively with ester chain length, while the β transition is unaffected and there is a γ transition at very low temperature which increases with ester size, see Fig. 6.14. Accordingly, the β transition is attributed to —COOR group rotation since it occurs at about the same temperature and increasing the length of the ester group has no effect on the position.

Secondary transitions may be important in determining the toughness of a polymer and Rehberg et al.[91] related the brittle–ductile transition temperature to the β transition from the change with the length of ester side group in the polyacrylate series. Accordingly these transitions are important in understanding mechanical properties.

Crystalline polymers show a more complex dynamic mechanical behaviour than do amorphous polymers. There is always considerable controversy surrounding the interpretation of these. Boyer[92] has reported that polyethylene and many other polymers exhibit a double glass transition characteristic of amorphous material relatively free from constraining crystallites and the upper one at $T_g(U)$ which is constrained and increases with crystallinity. Certainly it is well established[90,93] that the damping of polyethylene as a function of temperature exhibits distinct maxima whose intensities and positions vary markedly with crystallinity and degree of branching (see Fig. 6.15).

Dynamic mechanical analysis can also be measured by varying the frequency at a series of constant temperatures or by variation of the

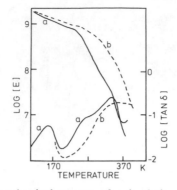

FIG. 6.15.   Dynamic mechanical spectra of polyethylenes; effect of branching.
Rheovibron—11 Hz. a, low density PE; b, linear high density PE. After P. J. Mills,
private communication.

temperature and using preset frequencies. This is particularly useful in
determining the change in the loss peaks with temperature and frequency,
giving a complete rheological description. This will enable the Williams,
Landel and Ferry master curves to be determined from the time–
temperature superimposition principle. Activation energies of the transitions are also determined from the change in temperature of the loss
peaks maxima with frequency. This is vital in assigning the nature of these
loss peaks.

The mechanical properties of nylon-1[94] have recently been studied using
a Rheovibron viscoelastometer and an Instron tensometer and the
conclusions compared with results from a DSC examination. Young's
modulus and tan δ were measured over a wide temperature range 11 Hz (see
Fig. 6.16) from which it was apparent that several transitions were present,
a β transition between 250–270 K and an α transition at 370 K. The
modulus fell slowly with temperature but approached a value similar to
rubbers near 370 K and so associated with the onset of the glass transition.
Differential scanning calorimetry measurements on the same samples
exhibited no transitions (α or β), and only a linear increase in the specific
heat with temperature. Nylon-1 is known to have an extended chain
conformation, and so the increase in number of conformations associated
with the onset of the glass transition would be small. The change in heat
capacity at the glass transition, $\Delta C_p(T_g)$ would be small and so undetect-
able by DSC.

Dynamic thermomechanical analysis of polymers, at different heating

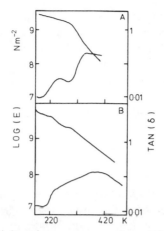

FIG. 6.16. Modulus and damping measurements on (A) poly-*n*-butyl isocyanate, (B) copolymer with ethyl isocyanate. Rheovibron—11 Hz.

rates and widely different frequencies, is extremely useful in detecting transitions and elucidating their nature. These transitions are frequently not readily observed by other techniques. However, the literature is extensive and there are excellent reviews on the subject[85,90,95] which list the wide range of application of these techniques and stress their potential in structural–properties relationships.

6.7. ELECTRICAL PROPERTIES

Thermoelectrometry includes all techniques in which an electrical characteristic of a material is examined as a function of temperature while programming the temperature. While there are many techniques which could be considered, only the two commonest will be dealt with—both associated with the apparent resistivity of the polymer insulators.

Insulators in general do not exhibit a high temperature dependence, and little change is observed except when the sample is actually undergoing a physical transition or a chemical reaction, particularly as the products of the reaction are semi-conducting. Measurement of current flowing between two platinum or gold electrodes embedded in the sample, i.e. 10^{-2} to 10^{-15} A on applying a constant voltage, is made as a function of temperature, care being taken to design the electrode contacts and leads to eliminate contact and surface resistance effects. Direct and alternating current resistance bridges have been used. Chiu[96] has used electrothermal analysis to study the thermal decomposition reactions in polyacrylonitrile which has been examined above by other techniques. He observed little change in conductivity associated with either the glass transition temperature (380 K) and with the thermal coloration reactions occurring from 450–520 K, which have been considered to produce conducting polyimine units. However, the exothermic reaction which occurs above 550–600 K, associated with the evolution of volatiles, also produced a large increase in conductivity. However, this must be associated with the reaction producing charge carrying entities since above this temperature range the conductivity again falls. Only above 1000 K does the conductivity of the residue increase substantially and the semi-conducting properties are also observed in the residue at room temperature. Obviously the elimination of nitrogen and other volatiles increases the length of the conducting sequences, such that a semi-conducting 'graphitic' residue is produced. Pope[97] has described a similar study on the dehydrochlorination of PVC and subsequent decomposition of the poly-ene residue.

Chiu[96] has also observed a very large change in the conductivity of poly-(ethylene oxide) on converting it from solid to liquid or vice-versa, and this has been used recently, to study the melting and isothermal crystallisation of a high molecular weight sample. Poly(ethylene oxide) contains substantial ionic catalyst residues and a relatively large equilibrium concentration of water and undoubtedly this behaviour is associated with ionic conduction. Interestingly the kinetics of crystallisation analysed by the Avrami equation were meaningless mechanistically but the rate parameters agreed well with those determined by DSC.

Electret formation[98] and trapped charge carriers in polymers have also been used widely to determine secondary transitions and determine their relaxation times and activation energies. Persistent polarisation in polymers can be induced in polymers by many means including cooling under an applied field, irradiation with electrons,[99] and corona discharge.[100] Thermally stimulated discharge currents (TSD) are then observed by reheating the polymer at a slow rate (1–3 K min^{-1}). The currents are due to the liberation of induced charges from the electrodes as the polarisation decays in the sample. The relaxation of the polarisation closely follows molecular motions in the polymer, i.e. dipole relaxations, and so contains characteristic activation energies and relaxation times.[101]

In most polymers thermally stimulated discharge current spectra exhibit maxima near the glass transition, melting points and secondary transitions. Vanderschueren[102] has exhibited the particular power of this technique to examine in detail the β transition in a wide range of polar polymers using partial polarisation to simplify the analysis of the spectra. This involves cooling the specimen at a slow rate, e.g. 1 K min^{-1}, and the electric field is not applied continuously but over several well defined temperature ranges—each separated by a short-circuit period. On subsequent re-heating a corresponding number of partial peaks characteristic of the overall distribution is observed. This decomposes the initial broad complex peak into several components which enable the distribution of relaxation times to be determined. In practice, the number of steps which can be isolated without overlapping one another are 3–4. The distributions of activation energies associated with the β transition were determined from the partial peaks initial slopes with temperature. A similar analysis has been made of the relaxation characteristics associated with the β transition in poly-(methyl methacrylate)[103] using dielectric relaxation theory which obey the time–temperature superposition principle.

The shift factor, activation energy, dielectric relaxation strength, and the distribution function of relaxation times were calculated. These were in excellent agreement with dynamic measurements.

These studies indicate the usefulness of electret studies on polymers for meaningful interpretation of the molecular movements associated with primary and secondary transitions. The wealth of studies currently being reported indicates the importance placed on such studies.

6.8. OPTICAL METHODS

Many optical properties vary substantially with temperature and structure of the polymer. In particular the refractive index, absorption and extinction coefficients at fixed wavelengths, and emission of light can obviously be used. There are thus many potential thermal analytical units based on optical characteristics which can be used to study polymers. This will, however, be restricted to two important techniques which have made considerable contributions to understanding polymer properties and behaviour—thermomicroscopy and thermoluminescence.

A hot-stage attachment for an optical microscope is perhaps one of the simplest, obvious and oldest thermal analytical procedures that has been devised to study materials. With the development of long distance working lenses, and micro furnaces thermostatted by temperature controllers, damage to optics is minimal. Obviously, melting points can be determined from the last trace of visual crystals, and crystalline regions can be directly measured—their number and radii determined as a function of time and temperature. The radial growth rates and nucleation characteristics of a crystallising polymer can be determined over a range of temperatures and melting and cooling procedures. Crystalline polymers are birefringent and the extent of depolarisation can be used to determine the relative extent of crystallinity automatically from a photocell read out, i.e. the normalised light intensities at zero time I_0, finally I_∞ and at time t, I_t, have been used to determine relative crystallinity X_t, i.e.

$$X_t = (I_t - I_0)/(I_\infty - I_0)$$

and so the extent of crystallinity followed with time for analysis by the Avrami equation. Although this procedure has been widely used[104-6] it has recently been re-evaluated[107] by comparing the rates of crystallisation of polypropylene, poly(butylene terephthalate) and poly(ethylene tereph-thalate) by depolarisation microscopy and DSC. The crystallisation rates determined by microscopy were substantially slower. The possibility that birefringence did not depend linearly on crystallinity was considered.

However, the importance of this technique must surely rest on the ability to evaluate the individual nucleation and growth rate characteristics, see Fig. 6.17. Kovacs et al.[108] have isolated the individual lamellar crystals which make up the normal spherulitive texture in crystalline poly(ethylene oxide), by a process of self-seeding, isothermal growth of the seeds and rapid quenching. In this, smaller crystals decorate the surface of the larger ones, grown isothermally, and make them visible in the matrix of similar refractive index. In this way, the growth rate of various faces of chain extended, and once, twice and multiple chain folded crystals can be determined over a range of temperatures and the relative thermodynamic stabilities of the individual crystal structures determined. The current theories of nucleation and growth of polymer crystals have to be modified considerably to account for these elegant observations. Similarly, using a u.v. light microscope on doped polymer samples, Billingham et al.[109] have investigated the rejection of u.v. absorbing impurities by a crystallising polymer. Assuming the rejection to be complete and impurities are restricted to the amorphous phase, the build up of impurities within a crystallising spherulite indicates that the average crystallinity of unit volume of spherulite decreases as it grows larger. This is of prime importance in crystallisation kinetic theory since, in particular, Avrami's basic model assumed constant density of crystallinity within the spherical crystallising unit. Obviously such procedures will enable correction parameters to be determined for this density term, and evaluate its time

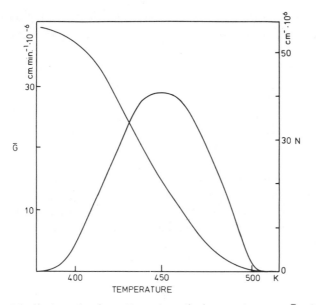

FIG. 6.17. Nucleation density, N, and radical growth rate, $\bar{G}$, during the isothermal crystallisation of isotactic polystyrene.

dependence. Theoretical calculations[110] have shown that this term will substantially alter the values of n in the Avrami equation.

Thermoluminescence is the measurement of total light emitted from a polymer as it is heated and it appears to be associated with the onset of mobility of trapped radicals within the polymer as the temperature of mechanical transitions in the polymer are exceeded.

Thermoluminescence has been observed in irradiated polyethylene, polytetrafluoroethylene, polycarbonate, poly(methyl methacrylate) and polystyrene[111] (see ref. 111 for subsequent reference to the other polymers). In the case of polystyrene, light emitted intensities reach maxima at 160, 220 and 376 K which were attributed to chain end carboxyl radicals, the cyclohexadienyl radical, and a main chain polystyryl radical decaying by radical–radical combination. This was substantiated by ESR measurements above the various temperature range. By a curve fitting procedure on the glow curves a second-order radical combination reaction mechanism was substantiated. X-ray irradiated PMMA exhibited only two peaks, at 140 and 370 K. ESR measurement indicated that the methyl methacrylate chain radical decay was responsible for the high temperature glow curve.

The low temperature glow curve appears to be associated with trapped electrons or small radicals which become mobile at γ transitions.[111]

6.9. CONCLUSIONS

Thermal analysis has made an impact on polymer science and will continue to do so with increasing effect. The various techniques have been used routinely to measure parameters for comparison, and as research tools they are successful in accumulating data rapidly. However, the data must be carefully and critically assessed with an appreciation of limitations of procedure.

There are, of course, many other physical and mechanical properties of polymers which have been studied as a function of temperature and time and which have not been listed. The review was not intended to be extensive. There will be further techniques which will become increasingly more important to the polymer scientist than those listed; we can only hope to recognise these as they become available. Thermomagnetometry, thermoacoustometry, thermoosonometry and others are well developed techniques with relevance to polymers, but they have not been included since, at the moment, they do not appear to be particularly main stream techniques. Neither has temperature drop turbidity, temperature jump techniques for measuring critical solution properties, or spectroscopic techniques such as i.r., broad line and narrow line NMR or EST been included since these techniques must be considered outside the terms of reference. This is, however, a matter of fine judgement since clearly some of these techniques must fulfil all the requirements of thermal analysis more so than many of the techniques listed above. The present author makes no apology for this since it reflects personal preferences. Hopefully, enough interesting examples have been included to suggest future trends in the subject, to encourage others into the field and to convey the vast amount of useful and informative studies being made on polymers by these techniques.

New developments are inevitable—one obviously expects an increasing tendency to interface the instruments to microprocessors to control the equipment, store base line corrections, and collect the data for further processing. Substantial improvements in instrument design and in electronics will also continue which will improve sensitivity and enable the polymer systems to be studied to slower rates and to further extents.

More than anything else these techniques will be applied to further polymer and copolymer systems, and comparison made between different techniques. This will inevitably lead to reconsideration of current theories on polymer structure–property relationships.

REFERENCES

1. G. Lombardi, *For Better Thermal Analysis*, 2nd edn., Published by the International Confederation for Thermal Analysis, Rome, 1980.
2. R. H. Still, *Brit. Polym. J.*, **11**, 101, 1979.
3. C. J. Keattch, *An Introduction to Thermogravimetry*, Heyden, London, 1969; R. F. Schwenker and P. D. Garn, *Thermal Analysis*, Vols. 1 and 2, Academic Press, New York, 1969.
4. I. Einhorn, *Thermal Analysis*, Polymer Conference Series, University of Utah, 1970.
5. R. C. Mackenzie, *Differential Thermal Analysis*, Vols. 1 and 2, Academic Press, London, 1970 and 1972, respectively.
6. T. Daniels, *Thermal Analysis*, Kogan Page, London, 1973.
7. A. Blazek, *Thermal Analysis*, Van Nostrand, Reinhold, 1974.
8. W. W. Wendlant, *Thermal Methods of Analysis*, 2nd edn., Wiley, New York, 1974.
9. B. Ke (Ed.), *Thermal Analysis of High Polymers*, Interscience, New York, 1974.
10. R. F. Schwenker, *Thermo-analysis of Fiber and Fiber Forming Polymers*, Interscience, New York, 1966.
11. H. Kambe and P. D. Garn, *Thermal Analysis*, Wiley, New York, 1974.
12. P. E. Slade and L. T. Jenkins (Eds.), *Techniques and Methods of Polymer Evaluation*, Vol. 1, Marcel Dekker, Inc., New York, 1966.
13. P. E. Slade and L. T. Jenkins, *Thermal Characterization Techniques*, Marcel Dekker, New York, 1966.
14. N. Grassie (Ed.), *Developments in Polymer Degradation—1*, Applied Science Publishers Ltd, London, 1977.
15. J. N. Hay, *Polymer*, **19**, 1224, 1978.
16. C. P. Bosnyak, *Ph.D. Thesis*, University of Birmingham, 1980.
17. J. R. MacCallum, *Brit. Polym. J.*, **11**, 120, 1979.
18. A. Barlow, R. S. Lehrle, J. C. Robb and D. Sunderland, *Polymer*, **8**, 537, 1967.
19. N. Grassie and H. W. Melville, *Proc. Roy. Soc.*, **199A**, 1, 1949.
20. I. C. McNeill, in *Developments in Polymer Degradation—1*, N. Grassie (Ed.), Applied Science Publishers Ltd, London, 1977, p. 43.
21. R. S. Lehrle, *Lab. Pract.*, **17**, 696, 1968.
22. R. S. Lehrle and J. C. Robb, *Nature*, **183**, 1671, 1959. A. Barlow, R. S. Lehrle, and J. C. Robb, *Polymer*, **2**, 27, 1961; *SCI Monograph No. 17*, 267, 1963.
23. F. A. Bell, R. S. Lehrle and J. C. Robb, *Polymer*, **12**, 579, 1971.
24. *a*. N. Grassie and J. N. Hay, *J. Polym. Sci.*, **56**, 189, 1962. *b*. J. N. Hay, *J. Polym. Sci.*, **6**, 2127, 1968.
25. N. Grassie and R. McGuchan, *Europ. Polym. J.*, **6**, 1277, 1970.
26. R. E. Peakman, *Ph.D. Thesis*, University of Birmingham, 1978; R. S. Lehrle and R. E. Peakman, private communication.

27. H. H. G. Jellinek, *J. Polym. Sci.*, **3**, 850, 1948; *ibid.*, **4**, 1 and 13, 1949; H. H. G. Jellinek and L. B. Spencer, *J. Polym. Sci.*, **8**, 573, 1952.
28. S. L. Madorsky and S. Strauss, *J. Res. Nat. Bur. Stand.*, **63A**, 261, 1959; S. L. Madorsky, D. McIntyre, J. H. O'Mara and S. Strauss, *J. Res. Nat. Bur. Stand.*, **66A**, 307, 1962.
29. N. Grassie and W. W. Kerr, *Trans. Faraday Soc.*, **53**, 234, 1957.
30. G. G. Cameron and N. Grassie, *Polymer*, **2**, 367, 1960.
31. G. G. Cameron and G. P. Kerr, *Europ. Polym. J.*, **4**, 709, 1968.
32. G. G. Cameron, *Makromol. Chem.*, **100**, 255, 1967.
33. J. R. MacCallum, *Makromol. Chem.*, **83**, 129, 1965.
34. G. Bagby, R. S. Lehrle and J. C. Robb, *Polymer*, **9**, 285, 1968.
35. C. D. Doyle, *Evaluation of Experimental Polymers*, WADD Technical Report 60-283 USAF, Wright Patterson AFB, Ohio, 1960.
36. G. F. Skala, *J. Rech. Atmos.*, 189, 1966; *Anal. Chem.*, **35**, 702, 1963.
37. C. B. Murphy and C. D. Doyle, *Appl. Polym. Symp.*, **2**, 77, 1966.
38. J. D. B. Smith, D. C. Phillips and T. D. Kaczmarek, *Anal. Chem.*, **48**, 1976.
39. F. W. Van Luik and R. E. Rippere, *Anal. Chem.*, **34**, 1617, 1962,
40. A. F. Dillon, R. S. Lehrle, J. C. Robb and D. W. Thomas, *Adv. in Mass Spectroscopy*, **4**, 89, 1967.
41. S. L. Boersma, *J. Amer. Ceramic Soc.*, **38**, 281, 1955.
42. M. J. Richardson and N. G. Savill, *Polymer*, **18**, 413, 1977.
43. M. J. Richardson and N. G. Savill, *Brit. Polym. J.*, **11**, 123, 1979.
44. M. A. De Bolt, A. J. Easteal, P. B. Macedo and C. T. Moynihan, *J. Amer. Ceram. Soc.*, **59**, 16, 1976.
45. T. G. Fox and P. J. Flory, *J. Appl. Phys.*, **21**, 581, 1950.
46. L. C. E. Struik, *The Physical Ageing of Amorphous Polymers and Other Materials*, Elsevier, Amsterdam, 1978.
47. G. Adams, J. N. Hay, R. N. Haward and I. W. Parsons, *Polymer*, **17**, 51, 1976.
48. A. A. Aref-Azar, *Ph.D. Thesis*, University of Birmingham, 1980.
49. G. Tammann, *Z. Anorg. Chem.*, **110**, 166, 1920.
50. P. J. Flory, *J. Chem. Phys.*, **17**, 223, 1949.
51. J. D. Hoffmann and J. J. Weeks, *J. Res. Nat. Bur. Stand.*, **64A**, 73, 1960; *ibid.*, **65A**, 297, 1961.
52. L. Mandelkern and M. J. Gopalan, *J. Phys. Chem.*, **71**, 3833, 1967.
53. A. M. Afifi-Effat, *Ph.D. Thesis*, University of Birmingham, 1972.
54. J. N. Hay, *J. Polym. Sci., Phys. Ed.*, **14**, 2845, 1976.
55. P. J. Flory and A. Vrij, *J. Amer. Chem. Soc.*, **85**, 3548, 1963.
56. C. Booth, D. R. Beech, D. U. Dodgson, R. K. Sharpe and J. S. Waring, *Polymer*, **13**, 73 and 246, 1972.
57. G. R. Slater, *M.Sc. Thesis*, University of Birmingham, 1979.
58. C. Ibbotson and R. P. Sheldon, *Brit. Polym. J.*, **11**, 146, 1979.
59. A. Ziabichi, *Colloid Polym. Sci.*, **252**, 435, 1974.
60. J. N. Hay, *Brit. Polym. J.*, **11**, 137, 1979.
61. A. A. Aref-Azar, J. N. Hay, B. J. Marsden and N. Walker, *J. Polym. Sci., Phys. Ed.*, **18**, 637, 1980.
62. M. Wiles and J. N. Hay, *J. Polym. Sci., Phys. Ed.*, **17**, 2223, 1979.
63. J. M. Barton, *Brit. Polym. J.*, **11**, 115, 1979.

64. S. Sounour and M. R. Kamal, *Thermochim. Acta*, **14**, 41, 1976.
65. J. P. Kennedy and C. M. Fontana, *J. Polym. Sci.*, **39**, 501, 1959.
66. E. V. Thompson, *J. Polym. Sci., B*, **4**, 361, 1966.
67. B. Wunderlich and M. Dole, *J. Polym. Sci.*, **24**, 201, 1957.
68. F. H. Müller, *Rubber Chem. Tech.*, **30**, 1027, 1957; *Pure and Appl. Chem.*, **23**, 201, 1970.
69. G. P. Andrianova, B. A. Arutyunou and Y. U. Popou, *J. Polym. Sci., Phys. Ed.*, **16**, 1139, 1978.
70. J. W. Maher, R. N. Haward and J. N. Hay, *J. Polym. Sci., Phys. Ed.*, **18**, 2169. 1981.
71. J. W. Maher and J. N. Hay, private communication.
72. J. N. Hay, *J. Polym. Sci.*, **3A**, 433, 1965.
73. A. M. Afifi-Effat and J. N. Hay, *Brit. Polym. J.*, **9**, 290, 1977.
74. W. Banks, M. Gordon, R. J Roe and A. Sharples, *Polymer*, **4**, 61, 1963.
75. A. Booth and J. N. Hay, *Polymer*, **10**, 95, 1969.
76. J. N. Hay, M. Wiles and P. A. Fitzgerald, *Polymer*, **17**, 1015, 1976.
77. P. Meares, *Polymers—Structure and Properties*, Van Nostrand, London, 1964.
78. J. N. Hay, *Brit. Polym. J.*, **3**, 74, 1971.
79. P. C. Ashman, C. Booth, D. R. Cooper and C. Price, *Polymer*, **16**, 899, 1975.
80. J. N. Hay, *Brit. Polym. J.*, **9**, 72, 1977.
81. R. A. Haldon and R. Simha, *Bull. Amer. Physic. Soc.*, **12**, 368, 1967; R. A. Haldon, W. J. Schell and R. Simha, in *Cryogenic Properties of Polymers*, T. T. Serafini and J. L. Koenig (Eds.), Marcel Dekker, Inc., New York, 1968, p. 137.
82. R. N. Haward, J. N. Hay and N. Walker, *J. Mat. Sci.*, **14**, 1085, 1979.
83. I. W. Gilmour, A. Trainor and R. N. Haward, *J. Polym. Sci., Phys. Ed.*, **16**, 1277, 1978.
84. I. W. Gilmour, A. Trainor and R. N. Haward, *J. Polym. Sci., Phys. Ed.*, **16**, 1291, 1978.
85. L. E. Nielsen, *Mechanical Properties of Polymers*, Reinhold Publishing Co., New York, 1962.
86. J. K. Gillham, in *Techniques and Methods of Polymer Evaluation*, Vol. 2, *Thermal Characterization Techniques*, P. E. Slade and L. T. Jenkins (Eds.), Marcel Dekker, Inc., New York, 1970.
87. J. K. Gillham and R. F. Schwenker, *Appl. Polym. Symp.*, **2**, 59, 1966.
88. S. Strella, *ASTM Bull.*, **214**, 47, 1956; D. J. Robinson, *J. Sci. Instrum.*, **32**, 2, 1955; A. W. Nolle, *J. Appl. Physics*, **19**, 753, 1948; S. Newman, *J. Appl. Polym. Sci.*, **2**, 333, 1959.
89. B. Maxwell, *ASTM Bull.*, **215**, 16, 1956; *J. Polym. Sci.*, **20**, 551, 1956.
90. J. Heijboer, *Brit. Polym. J.*, **1**, 3, 1969.
91. C. E. Rehberg, M. B. Dixon and W. A. Faucette, *J. Amer. Chem. Soc.*, **72**, 4307 and 5199, 1950.
92. R. F. Boyer, *Macromolecules*, **6**, 288, 1973.
93. D. E. Kline, J. A. Sauer and A. E. Woodward, *J. Polym. Sci.*, **22**, 445, 1956.
94. A. A. Owadh, I. W. Parsons, J. N. Hay and R. N. Haward, *Polymer*, **19**, 386, 1978.
95. R. F. Boyer, *J. Polym. Sci., Polym. Symp.*, 189, 1975.

96. J. Chiu, *Anal. Chem.*, **39**, 861, 1967.
97. M. I. Pope, *Polymer*, **8**, 49, 1967.
98. J. Turnhout, *Thermally Stimulated Discharge of Polymer Electrets*, Elsevier, New York, 1975.
99. P. K. C. Pillai and Rashmi, *J. Polym. Sci., Phys. Ed.*, **17**, 1731, 1979.
100. G. M. Sessler and J. E. West, *Appl. Phys. Letters*, **17**, 507, 1970.
101. R. A. Cresswell and M. M. Perlman, *J. Appl. Phys.*, **41**, 2365, 1970.
102. J. Vanderschueren, *J. Polym. Sci., Phys. Ed.*, **15**, 873, 1977.
103. Ch. A. Solunov and Ch. S. Ponevsky, *J. Polym. Sci., Phys. Ed.*, **15**, 969, 1977.
104. I. H. Magill, *Polymer*, **3**, 25, 1962.
105. F. L. Bingsbergen and B. G. M. deLange, *Polymer*, **11**, 309, 1970.
106. J. B. Jackson and G. W. Longman, *Polymer*, **10**, 873, 1969.
107. C. F. Pratt and S. Y. Hobbs, *Polymer*, **17**, 12, 1976.
108. A. J. Kovacs, C. Straupe and A. Gonthier, *J. Polym. Sci., C*, **59**, 31, 1977.
109. N. C. Billingham, P. D. Calvert, J. B. Knight and T. G. Ryan, *Brit. Polym. J.*, **11**, 155, 1979.
110. J. N. Hay and Z. J. Przekop, *J. Polym. Sci., Phys. Ed.*, **17**, 951, 1978.
111. S. Radhakrishna and M. R. K. Murthy, *J. Polym. Sci., Phys. Ed.*, **15**, 1261, 1977.
112. S. Radhakrishna and M. R. K. Murthy, *J. Polym. Sci., Phys. Ed.*, **15**, 987, 1977.

Chapter 7

EBULLIOSCOPIC METHODS FOR MOLECULAR WEIGHTS

G. DAVISON

Chorley, Lancashire, UK

7.1. INTRODUCTION

7.1.1. The Ebullioscopic Method

When a non-volatile solute is dissolved in a liquid, the vapour pressure of
the liquid is lowered. This phenomenon has been known qualitatively for
nearly two hundred years and was even studied quantitatively at the
beginning of the 19th century by Bethellet (1803) and Faraday (1822).

A direct consequence of the reduction of vapour pressure by a non-
volatile solvent is that the boiling point of the solution, i.e. the temperature
at which the vapour pressure is equal to the atmospheric pressure, must be
higher than that of the pure solvent. Although studies of the rise in boiling
point produced by dissolved substances were made by several workers (in
1887 van't Hoff[1] deduced a relationship between the elevation of the boiling
point and the molecular weight of the solute), it is to Beckman (1889)[2] that
most of the credit must be given for accurate measurements and for the
experimental procedure which is to some extent the basis of the modern
methods. The experimental procedure is described as the 'Ebullioscopic
Method'.

It is of interest that a number of fundamental problems which beset
Beckman in his work are still the source of major problems. In his method,
the thermometer was placed directly in the liquid. Since the temperature of
the boiling liquid is dependent upon pressure, the position of the
temperature sensor is of great importance, as the temperature recorded for
the boiling liquid is dependent upon the depth of immersion of the sensor.
Although there have been many developments in temperature sensing
devices, the actual point of measurement is still of great importance.

Variations in atmospheric pressure may also lead to errors in the
experimental determination of molecular weight using this technique.

209

However, the difficulty associated with this pressure dependence of temperature measurement has been overcome in part by the use of dual vessel ebullioscopes—supposedly identical apparatus—in which one system contains pure solvent and the other receives the additions of weighed solute. An added advantage of this approach is that the problem of 'solvent hold-up' can be overcome to some extent. This arises because in an ebullioscopic apparatus a proportion of the solvent will be contained in the reflux condenser as pure solvent. As a consequence, the amount of solvent in the solution is reduced and the solute concentration is increased.

In this context Billingham[3] has pointed out that most high sensitivity ebullioscopes use some form of differential measurement to enable the boiling point of the solution to be compared to that of the pure solvent at the same pressure. Further it is noted that since the change in ebullioscopic constant is usually only about 0.3% for a 10 mm change in pressure, the problem can be neglected for measurements on low molar mass solutes. However, in a high sensitivity ebullioscope, atmospheric pressure fluctuations can appear as noise on the output of the temperature sensor and some form of pressure stabilisation is therefore necessary.

Another major problem is that of superheating, i.e. when the temperature of the liquid is above the true boiling point. This is due to the fact that because of hydrostatic pressure, the pressure at a point below the surface of the liquid is greater than atmospheric pressure. Hence the liquid temperature at this point may be insufficient to cause the phenomenon of boiling because, in any bubble formed below the surface, it is necessary to generate a pressure above atmospheric to compensate for the effect of the curvature of the bubble. It is also necessary that the heater element and boiling surfaces are at a temperature greater than the true boiling point of the liquid in order to boil the liquid.

A device which is known as a Cottrell pump[4] (Fig. 7.1) is a means whereby vapour bubbles lift the boiling solvent from the solution over a measuring thermistor. The object here is to allow superheating to be lost before the solution reaches the thermistor. Even though a number of devices such as the Cottrell pump have overcome the problem to some extent, it is a problem which largely remains unsolved.

Apart from the experimental difficulties already described there are a number of other situations which lead to particular problems in operating this technique.

Perhaps the most intractable one it is necessary to take account of, arises particularly in determinations of number average molecular weight on polymer samples and is *foaming*.

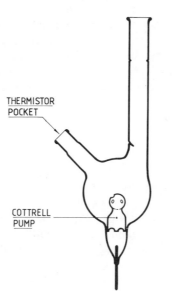

THERMISTOR
POCKET

COTTRELL
PUMP

FIG. 7.1.   Ebulliometer showing position of Cottrell pump (Gallenkamp type).

Normally the Cottrell pump, utilising boiling pure solvent or a non-foaming solution enables an intimate mixture of liquid and bubbles to be cascaded over the temperature sensing element.

In the case of a polymer solution however, an increasing ratio of foam to liquid is produced by the Cottrell pump and beyond a certain concentration of sample, only foam is produced. The consequence is that the temperature sensing device only receives decayed foam which tends to contain a higher concentration of solute than that present in the bulk of the liquid.

Various attempts have been made to reduce the effect of foaming in polymer solutions. These have often and more recently involved the use of 'rotating' or 'rocking' ebullioscopes, as described later, but the difficulties which arise from foaming in polymer solutions have yet to be fully overcome.

The elevation of boiling point is seldom measured directly in temperature units. Much more commonly a comparison is made between the temperature sensing device response for the solvent and for the solution of polymer. This may be achieved within a single suitably constructed ebullioscope or by utilising twin ebullioscopes. In both these approaches the difference in detector response is determined as an off-balance bridge output in units of resistance which is proportional to the elevation of temperature.

Even in these circumstances it is still necessary to be able to determine the true concentration of polymer in the bulk of the liquid. It is clear that this concentration will be affected by what is known as the 'solvent hold-up', referred to previously.

In order to overcome these problems it has become much more common to use a 'calibration method' in which the ideal ebullioscopic constant and the solvent hold-up and other variables are included in an apparatus or ebullioscopic constant.

When such a system employs dual vessel ebullioscopes, variations in atmospheric pressure may also be accounted for.

The most commonly adopted experimental procedure is that which involves a series of preliminary experiments using pure compounds of accurately known molecular weight in order to obtain graphs of detector response against solute concentration (known as the calibration graph). Experiments are then carried out to relate detector response for polymer solutions over a range of concentrations.

Using the previously obtained calibration graphs it is then possible to determine the polymer of unknown molecular weight.

Thus, with non-polymeric solutions a simple calculation procedure comparing compounds of known and unknown molecular weight is all that is required.

However, the behaviour of most polymeric and some non-polymeric materials deviates from ideality to such an extent that special treatment of the data is necessary.

Examples of such a statistical approach to the examination of experimental data can be seen in the application of both the zero point and the divided difference methods of calculation described later.

7.1.2. Number Average Molecular Weights and the Sensitivity of the Technique

In reviewing some of the more recent work carried out using the technique of ebullioscopy to determine number average molecular weights an examination of the literature reveals that there are two main areas where most attention has been given to the development of the method.

First, the comparatively small number of experimenters using these ebullioscopic procedures have concentrated on devising apparatus which overcomes, to a greater extent, the problem of superheating, either by altering the geometry of the apparatus or by the incorporation of mechanical pumps to pump the boiling solvent/vapour over the tip of the temperature sensing device.

Second, more attention has been directed towards the interpretation of data obtained in experimental work, particularly in connecting with the possible molecular weight dependence of the calibration constant and the examination of anomalous behaviour often observed with polymer solutions.

Since the magnitude of the change in colligative property is inversely proportional to the molecular weight of the solute, the examination of relatively dilute solutions of polymers in these experiments leads to quite small changes in colligative property. For example, Moore and Tidswell[5] have shown that for a 1 % w/v solution of a polymer of $\bar{M}_n = 50\,000$ in a typical organic solvent, the magnitude of the elevation of boiling point is of the order of $5 \times 10^{-4}\,°C$.

Therefore, in order to measure the very small elevations of boiling point which occur when a high molecular weight solute is used, it is usual to employ either sensitive thermocouples or a pair of matched thermistors in a Wheatstone bridge circuit.

Many of the recent developments associated with this technique have been concerned with improvements in the apparatus used in the experiments. In particular, the use of more sensitive temperature detecting devices has enabled the range of this method to be extended considerably. Thus, Glover[6] has achieved a sensitivity of $1 \times 10^{-5}\,°C$ using a 160-junction thermocouple, and Zichy[7] has reported sensitivities of $2·4 \times 10^{-6}\,°C$ using thermistors as temperature sensing devices.

In general, however, the boiling point elevation is not measured directly as a temperature difference, but is taken from readings of the apparatus recording the temperature.

7.1.3. Range of Application of the Technique—Its Advantages and Limitations

In order to obtain the highest accuracy and precision, the most important experimental requirements in any ebullioscopic determination of molecular weight is a perfect equilibrium between vapour and liquid at the boiling point.

A number of recent improvements in the design of apparatus have been concerned with achieving this objective.

In the case of high molecular weight polymer samples the temperature changes being observed are quite small, e.g. $2 \times 10^{-5}\,°C$, and for some considerable time it was thought that the upper limit for the determination of $\bar{M}_n$ was in the region of 40 000. Thus, several workers including Ray,[8] Smith[9] and Glover and Stanley[10] have described determinations of molecular weight in the region 30 000–40 000 with good precision.

Nevertheless in a report by Zichy,[7] values of up to 100 000 were quoted with 90 % confidence limits of 10–15 % tolerance.

In reviewing previous work Glover and Kirn[11] pointed out that they were unaware of any previous reference to the use of methods for determining molecular weight above 100 000. At the same time the authors reported the use of an ebullioscope consisting of a conventional Cottrell pump with an 80-junction copper–constantan thermopile used in a differential manner as the temperature sensing element. Samples of polystyrene (one narrow, one wide distribution of molecular weight) were tested. Values of $\bar{M}_n$ for the narrow distribution samples obtained by ebullioscopic procedures showed good agreement with osmometry determinations; however, there was very considerable variation in this respect for the wide molecular weight distribution sample. This is possibly to be expected from fundamental differences between the techniques of ebullioscopy and osmometry.

With essentially the same apparatus some eight years later (1973), Glover[12] reported work involving 8000 determinations with a range of solvents and solutes. Most interesting was the use of a novel system, hexafluoro-2-propanol. This solvent has been used for a wide range of polymer samples, many of which are insoluble in other solvents at the concentration and temperature required.

Typical of some of the problems which can arise is that described in work carried out by Williamson[13] involving attempts to measure the number average molecular weight of high density polyethylenes by ebullioscopy. Using an ebullioscope similar to the design of Schon and Shulz,[14] the apparatus was calibrated against anthracene with bromobenzene as the solvent. The operation of the Schon and Schulz ebullioscope differs from conventional apparatus in that the liquid under examination is not pumped over the temperature sensing device. In this system temperature equilibrium is achieved by maintaining the temperature surrounding the liquid only slightly above its boiling point. Surface evaporation is then brought about by the vibration or shaking of the liquid. Good linear plots of elevation against concentration were obtained showing very small zero point error. However, when using high density polyethylenes, sigmoidal curves were obtained for the boiling point dependence on concentration. Although at high concentration this effect could have been attributed to the effects of foaming or the presence of superheating, the non-linearity at low concentrations could not be explained.

However, more recently other workers[12] have been able to demonstrate the successful determination of molecular weights ($\bar{M}_n$) up to 100 000 in good agreement with results obtained by other methods.

For instance, Parrini and Vacanti[15] have used a new ebullioscope operated in a differential mode to determine molecular weights of unfractionated isotactic polypropylenes. For temperature sensing they utilised a thermopile with 150-junctions and paid special attention to ensuring regularity of boiling by modification of the Cottrell pumps. As a result, it was found possible to determine molecular weights up to 85 000 with good reproducibility and an accuracy of 6·5 %. Of special importance in this work were several aspects of the experimental technique used. All solvents were freshly distilled and dried over metallic sodium. Several purified calibration compounds ranging in $\bar{M}_n$ from 390–891 were used and good agreement was observed between values determined for the apparatus constant. All samples were purified to remove stabilisers and compacted into pellets.

Most importantly, after every second M_n determination the ebullioscope was washed successively with chromic acid mixture, cold distilled water, acetone and three times with the solvent being used. The authors emphasised that washings before and after measurement must be made with the solvent being used in the determination.

As a result of this work it was concluded that whilst reproducibility of the measurements in the same solvent was fairly good there were considerable differences from solvent to solvent. Further, in order to obtain reproducible results, 'pre-conditioning' of the glass walls of the ebullioscope was essential. Despite the success of this work problems with foaming were experienced, the amount of foam decreasing on passing from chlorobenzene to ethylbenzene and to *m*-xylene as solvent.

Some problems associated with zero point error have been described by Rapoport and Taits[16] who investigated the factors affecting the accuracy of precision ebullioscopy. They proposed a method of calculation with corrections for atmospheric pressure and for the temperature dependence of the volume of solvent used. It was claimed that a graphical method of analysis of the elevation versus concentration plot had been developed in order to correct for the error of the first point (zero error). The problems associated with zero point error were ascribed to the fact that only a fraction of the first of the sample added to the solution was effective in raising the boiling point of the solution. This statement is misleading because it suggests that the zero error is associated only with the first point on the plot. Zero error arises when every point is low (or high) to the same extent.

Mention should also be made of a recent development in ebullioscope design which effectively extends the range for this technique and is claimed to overcome the problems associated with foaming. Sotobayashi et al.[17] have described the operation of a pressure-controlled twin ebullioscope in which a rotating unit is used instead of a Cottrell pump. It is possible with this equipment to choose the boiling temperature by setting the pressure with a pressure regulator.

Polystyrene and other polymers have been examined in this rotation ebullioscope without experiencing foaming.

Thus, it is claimed that providing sufficient care is taken to maintain good equilibrium between liquid and vapour and that adequate precautions are taken to purify solvent and solute, values of $\bar{M}_n$ up to and beyond 100 000 molecular weight should be capable of being determined by ebullioscopy.

However, very little if any of the equipment described is available commercially and models often vary widely in design.

7.1.4. Review of Ebullioscope Design
Although at the start of this survey a mention was made of the work of Beckmann, the use of a Beckmann thermometer is not satisfactory to determine the small temperature differences observed with high molecular weight polymers. Numerous advances in thermometer design took place before the introduction of electrical temperature sensing techniques utilising multi-junction thermocouples or thermistors.

A number of reviews have described the various types of ebullioscopes and their advantages and disadvantages. For instance, Glover[6] has categorised ebullioscopes as falling into three types, and these are described below.

The *simple* type of ebullioscope is represented by the early Beckmann systems where the ebullioscope is used to measure the temperature of a boiling liquid usually after it has been raised with a vapour lift pump and allowed to flow in a thin film over the temperature sensor.

In contrast to the simple ebullioscope the *differential* ebullioscope provides for instantaneous measurement of the difference in temperatures of the boiling liquid and its condensing vapour. This system has been used in two modes. In the first mode a Menzies and Wright[18] arrangement involved the sensing of the temperature of the boiling liquid by the lower bulb of a differential vapour pressure thermometer, whereas the temperature of the condensing vapours was sensed by the upper bulb. In the second mode or independent differential ebullioscope there is independent measurement of the temperatures of boiling liquid and condensing vapours. As Glover[6] points out, by independent measurement of these temperatures, changes in the vapour condensation or reference temperature are easily detected.

The third system which may be utilised is the *twin or paired* arrangement of ebullioscopes. The ebullioscopes are placed side by side with a simultaneous, independent measurement of the boiling temperature of pure solvent in one ebullioscope and that of the solution in the other. Swietoslawski and Anderson[19] recommend the use of this system as the best arrangement for the determination of molecular weights.

The system reported by Schon and Schulz[14] differs from other types of ebullioscope in that the liquid under test is not pumped over the temperature sensor; surface evaporation is induced by shaking. A more recent modification of the system by Ezrin[20] involves the temperature sensors being placed just above the liquid surface and being bathed by a spray of liquid initiated by vibration. Lehrle,[21] in a survey of ebullioscopes in use up to 1961, noted that the principal advantage of differential instruments is that both temperature elements are affected by ambient conditions in a similar manner. It is evident that this is more difficult to ensure in a twin ebullioscope. However, Lehrle observes that twin instruments may be simpler to construct and maintain and have the further advantage that the boiling points of both solvent and solution are sensed by the same technique but this may lead to a higher noise. A point not noted by Lehrle was the much greater cost of such a system.

Most of the ebullioscopes designed for polymer work are of the differential type rather than the twin system but in at least one case[22] an apparatus has been used incorporating the principles of both types of ebullioscope in one instrument.

Lehrle and Majury[23] devised a two thermistor ebullioscope in which the zero reading corresponding to pure solvent could be monitored throughout the experiment. It was noted by Daniels[24] that the main disadvantage of the usual types of ebullioscope is that the boiling point of the pure solvent is determined at the beginning of the experiment with pure solvent and assumed to remain constant throughout the experiment. The apparatus of Lehrle and Majury suffered from two main disadvantages both connected with the movement of the central unit which contained the two thermistors: (a) it was difficult to locate the central unit reproducibly and hence the actual geometry varied during the run, and (b) thermal equilibrium was disturbed by this movement. As a result the reproducibility of the readings obtained was not good.

An instrument of similar design was described by Lehrle and co-workers;[24] its central unit was fixed and the apparatus contained a matched triplet of thermistors. Two of the thermistors ('vapour' ones) were located in the vapour region above the boiling liquid and the third ('liquid' thermistor) received effluent directly from the Cottrell pump. With this ebullioscope the reproducibility was improved and the speed of measurement was also increased. The glassware of the instrument was similar to that of the Lehrle and Majury instrument but a glass 'umbrella' was included above the nozzles of the Cottrell pump, the purpose of this was to prevent droplets of returning solvent reaching the region of the lower thermistor.

Following on this work De Pablo[25] made several modifications on Foster's instrument (including re-siting of the thermistors), and as a result optimum values for the heater voltage and solvent volume were determined. More importantly the advantages of using one, two and three thermistor instruments were assessed and he concluded that the three thermistor design was the one most suited to polymer studies. The reason for this was that the three thermistor instrument is free from moving parts and provided that the thermistors were well matched, only a small drift in the outputs as a result of atmospheric pressure changes is observed.

Daniels[24] further developed the De Pablo recommended system and discussed the use of thermistors to measure temperature, the circuits and the importance of thermistor matching in ebullioscopy. In order to follow previously successful determinations on low molecular weight compounds with work on polymers, Daniels investigated the source of noise in the ebullioscope. Further improvement was obtained after construction of, in effect, a new ebullioscope without a vapour jacket but retaining a vacuum jacket, and the inner surfaces of the latter were silvered (except in the region of the observation window). The external insulation (vermiculite) was repacked around the ebullioscope in order to eliminate air spaces where convection currents could possibly develop.

Daniels reported that he found the most stable output was obtained when the Cottrell pump was only just pumping since as the rate of pumping increased so did the noise. It was found that for polymer solutions it was necessary to use a four-armed pump to remove the large volume of foam which collected in the bell of the pump. In this system the ebullioscope was washed out with fresh solvent between each pair of determinations.

Daniel's work with the modified ebullioscope involved determinations of $\bar{M}_n$ on two polymer samples described as 'polyoxylated propylated glycerols of molecular weight 1000 and 3000'. Use of benzil (MW, 210·23) as calibrant gave him results which agreed with the determination of $\bar{M}_n$ by end-group analysis. Work was carried out by investigating situations of high concentration (84 g/kg) and low concentration (0–10 g/kg). (The former showed evidence of non-ideality as expected.)

Further determinations of $\bar{M}_n$ were carried out on samples of polystyrene ($\bar{M}_n$, 11 800; 15 000; 28 000; 34 000) in benzene and for one sample in cyclohexane. Good agreement was observed for the sample of $\bar{M}_n$ 11 800, in other cases the presence of 'S' shaped (sigmoid) curves[26] was observed.

In an extension of the work he observed that determinations on one of the polymer samples showed that less foam was produced at relatively low heater settings. He found, however, that in order to operate a Cottrell pump, a minimum heating rate was required which was always greater than the heating rate necessary just to boil the liquid. He thus concluded that extra foaming and superheating are to be expected from Cottrell pump instruments. He therefore constructed an ebullioscope incorporating a mechanical reciprocating pump using a three thermistor measuring system. Original designs of this equipment involved a 'one-arm pump', which however was superseded by a three-armed pump. However, in the final arrangement of the apparatus a four-arm mechanical pump was employed. He concluded that although the amount of 'S' shaped (sigmoid) curvature in the data was reduced when the mechanically pumped ebulliometer was used, it was not completely eliminated.

Glover[12] has described an ebullioscopic system for routine and special determinations of molecular weights which utilises a simple ebullioscope, a platinum immersion heater and a Cotterell type pump. Temperature sensing is by differential thermopile and the precision obtained varies between 1 and 6 %. Glover claims that thermopiles have advantages in that they are very stable in operation and change little, if at all, with age. It is further claimed that one particular advantage of thermopiles is the levelling effect of the multiple sensing points on the rapid temperature fluctuation caused by uneven boiling or pumping. (The present author attributes this to the relatively low speed of response of the multi-junction thermocouple.) The apparatus described, unlike that employed by Daniels, incorporated a vapour jacket. In contrast to some previous work a decreasing heat input programme was followed to eliminate the effects of both superheating and foaming.

One problem experienced with the Cottrell pump used was that the rate of pumping and heat input to the boiling solution could not be varied independently. Therefore, to overcome the problems of superheating, a no-dead-space mechanical pump was designed and used. It was indicated that further work would involve an examination of the operation of a dual vessel ebullioscope based on the instrument described. Also, provision had been made in the apparatus for samples of solution to be removed for analysis in order to investigate the distribution of solvent and solute in the apparatus.

7.2. BOILING POINT ELEVATION BEHAVIOUR

7.2.1. General Relationships

Although relationships between elevation of the boiling point and molecular weight have been derived by, for instance, Lehrle, some account must be taken of the activity of the solute and solvent. For 'ideal' dilute solutions the activity can be equated to molar concentration and the relationship can even be used in non-ideal solutions as an approximation. Most polymer solutions exhibit non-ideality and the problem is to arrive at a relationship which enables the dependence of the activity of the solvent on the concentration to be determined quantitatively. In the situation where a pure solvent is boiling with the vapour and liquid in equilibrium, the chemical potential of the solvent in the vapour phase and the liquid phase will be equal, i.e.

$$\mu_1(T) = \mu_1^v(T) \tag{7.1}$$

where μ_1 and μ_1^v are the chemical potentials of the solvent in solution and in the pure vapour, respectively.

As Lehrle[21] shows, if the chemical potential of pure solvent is represented by μ_1^0 then $\mu_1(T)$ may be replaced by $\mu_1^0(T) + RT \ln a_1$ where a_1 is the activity of the solvent in a solution at temperature, T. Therefore

$$\mu_1^0(T) + RT \ln a_1 = \mu_1^v(T) \tag{7.2}$$

or

$$R \ln a_1 = \frac{\mu_1^v(T)}{T} - \frac{\mu_1^0(T)}{T} \tag{7.3}$$

For a pure solvent at its boiling point, T_0, then:

$$\mu_1^0(T_0) = \mu_1^v(T_0) \tag{7.4}$$

Thus

$$\frac{\mu_1^0(T_0)}{T_0} = \frac{\mu_1^v(T_0)}{T_0} \tag{7.5}$$

If the solution boils at $(T_0 + \theta)$ where θ is the elevation of the boiling point, then combining eqns. (7.3) and (7.5):

$$R \ln a_1 = \left[\frac{\mu_1^v(T_0 + \theta)}{T} - \frac{\mu_1^v(T_0)}{T_0} \right] - \left[\frac{\mu_1^0(T)}{T} - \frac{\mu_1^0(T_0)}{T_0} \right] \qquad (7.6)$$

Then

$$R \ln a_1 = \int_{T_0}^{(T_0 + \theta)} \frac{\partial(\mu_1^v/T)}{\partial T} \, dT - \int_{T_0}^{(T_0 + \theta)} \frac{\partial(\mu_1^0/T)}{\partial T} \, dT \qquad (7.7)$$

$$= \int_{T_0}^{(T_0 + \theta)} \frac{(H_1^0 - H_1^v)}{T^2} \, dT \qquad (7.8)$$

where $(H_1^0 - H_1^v)$ is the difference in molar heat content between pure liquid and pure vapour, which is of course of the same magnitude (but conventionally not of the same sign) as the molar heat of evaporation (ΔH_e). Hence

$$R \ln a_1 = - \int_{T_0}^{T} \frac{\Delta H_e}{T^2} \, dT \qquad (7.9)$$

As ΔH_e is temperature dependent this expression can be rewritten:

$$\Delta H_e = \Delta H_0 + \Delta C_p(T - T_0) \qquad (7.10)$$

where ΔH_0 is the molar latent heat of evaporation of pure solvent at the boiling point and ΔC_p is the difference between molar heat capacity (specific heat) of the vapour and that of the liquid. If ΔC_p is assumed to be constant over the small concentration and temperature ranges, then:

$$RT \ln a_1 = - \int_{T_0}^{(T_0 + \theta)} \frac{\Delta H_0}{T^2} \, dT = - \int_{T_0}^{T_0 + \theta} \frac{\Delta C_p(T - T_0)}{T^2} \, dT \qquad (7.11)$$

Integration (assuming ΔC_p is constant) followed by expansion of the binomial and logarithmic terms gives:

$$\ln a_1 = - \frac{\Delta H_0 \theta}{RT^2} \left[1 - \left(\frac{1}{T_0} - \frac{\Delta C_p}{2\Delta H_0} \right) \theta + \left(\frac{1}{T_0^2} - \frac{2\Delta C_p}{3T_0 \Delta H} \right) \theta^2 \right] \qquad (7.12)$$

which is a general expression for the relationship between solvent activity and boiling point elevation. Lehrle notes that if θ is very small, powers higher than the first in θ may be neglected. Therefore,

$$\ln a_1 = - \frac{\Delta H_0 \theta}{RT^2} \qquad (7.13)$$

This approximation is equivalent to assuming constancy of ΔH_e over the boiling point elevation range, together with the assumption $T \times T_0 = T_0^2$ as can be seen by integrating eqn. (7.9) (ΔH_e constant) to give:

$$\ln a_1 = \frac{\Delta H_e}{R}\left(\frac{1}{T} - \frac{1}{T_0}\right) \tag{7.14}$$

which becomes eqn. (7.13) if we write $T \times T_0 = T_0^2$.

Lehrle suggests that for most polymer solutions, θ will be sufficiently small for the elevation eqn. (7.13) to be generally applicable. However, for very high concentrations and larger elevations higher terms in θ may have to be taken into account.

It has thus been shown that the activity of the *solvent* in solution is related, as above, to the elevation of boiling point. However, we wish to relate to the molal concentration of the solute to the elevation. It thus becomes necessary to consider how the activity of the solvent is related to the molal concentration of the solute. In order to do this an understanding of the non-ideality of the solution is required, i.e. we must propose a model for the non-ideality which permits the activity to be related to concentration. For dilute solutions (say less than 1 % w/v) of low molecular weight solutes, the model of ideal solutions may be applicable. For polymer solutions where non-ideality is likely to be present another approach is required.

For ideal solutions—where according to Raoult's law, the activity of each component is equal to its mole fraction—the activity of the solvent, a_1, is related to the mole fraction of the solute by the expression:

$$\ln a_1 = \ln x_1 = \ln(1 - x_2) \tag{7.15}$$

In dilute solution x_2 is small and we can approximate by neglecting second and higher terms in the expansion of $\ln(1 - x_2)$, i.e. $\ln(1 - x_2)$ is approximately equal to x_2.

Furthermore:

$$x_2 = \frac{\dfrac{W_2}{M_2}}{\dfrac{W_1}{M_1} + \dfrac{W_2}{M_2}}$$

where W_1 and W_2 are the mass of solvent and solute, respectively, and M_1 and M_2 are the molecular weights of solvent and solute, respectively. For dilute solution, and especially if M_2 is large, then W_2/M_2 is very small compared with W_1/M_1, hence:

$$x_2 = \frac{W_2}{W_1} \times \frac{M_1}{M_2} \tag{7.16}$$

Thus, $\ln a_1$ in the elevation eqn. (7.13) can now be replaced:

$$\frac{W_2 M_1}{W_1 M_2} = \frac{\Delta H_e \, \Delta T}{RT^2} \tag{7.17}$$

which on rearrangement gives:

$$\Delta T = \underbrace{\frac{RT^2 M_1}{1000 \, \Delta H_e}}_{(a)} \times \underbrace{\frac{1000 W_2}{W_1}}_{(b)} \times \frac{1}{M_2} \tag{7.18}$$

where (a) is a constant dependent only on the solvent—called the ebullioscopic constant (K_B) and (b) represents the concentration of the solution (C_0), in grams of solute per kilogram of solvent.

From this equation it is seen that a linear relationship exists between the elevation of boiling point, ΔT, and the concentration of solute, C_0, for ideal solutions. Therefore

$$\Delta T = \frac{K_B}{M} C_0 \tag{7.19}$$

7.2.2. Polymer Solutions and Non-ideality

As explained previously the above expression is inadequate for non-ideality in polymer solutions.

McMillan and Mayer[27] have shown that a power series will represent the relationship of elevation ΔT, to concentration thus:

$$\Delta T = A_1^* C_0 + A_2^* C_0^2 + A_3^* C_0^3 \tag{7.20}$$

where A_1^*, A_2^* and A_3^* are the 1st, 2nd and 3rd virial coefficients. Here A_1^* is equal to $K_{B/M}$ and is the first-order term for ideal solutions of concentration C_0. A_2^* and A_3^* depend mainly on the excess entropy of mixing and the excess heat of mixing of the polymer and solvent at the boiling point. Therefore the determination of the molecular weight of the solute requires the evaluation of the first virial coefficient, and the problem resolves into one of making due allowance for the effects of the second (and perhaps higher) coefficients.

The methods by which the data obtained can be interpreted to obtain a value for the molecular weight of a solute are discussed later.

7.2.3. The Ebullioscopic Constant

The ebullioscopic constant, as defined by eqn. (7.18) may be calculated from known parameters characteristic of the solvent.

In the 'absolute' method of determining the molecular weight by ebullioscopy this value of the ebullioscopic constant is used and it is necessary to determine the true concentration of the solution in the ebullioscope, i.e. by making a correction for solvent hold-up, and also to determine the elevation of boiling point in true temperature units. This approach is seldom used. Almost always a calibration method is employed in which the ideal ebullioscopic constant, K_B, the conversion factor to relate parameter of measurement to elevation in temperature units and the solvent hold-up are included in an apparatus constant, K_A. The value of K_A is then determined from a calibration experiment using a solute of known molecular weight.

Glover has emphasised that considerable care is still necessary, even in routine operation, to overcome problems caused by a number of variables in the system. These include problems with superheating, difficulty in knowing the true concentration of solution at the point of temperature measurement, and possible variation of ebullioscopic constant with solute molecular weight. This latter point, the molecular weight dependence of the calibration constant in ebullioscopy, is still the subject of discussion.[24]

Some aspects of the interpretation of experimental data, especially in relation to the possible molecular weight dependence of the calibration constant in ebullioscopy, have been reported. Glover and Hill[28] carried out an investigation involving a critical consideration of the ebullioscopic constant, K_B. The constant was measured in several of the more common solvents used for molecular weight determination, e.g. acetone, ethanol, benzene, carbon tetrachloride, etc. In this investigation boiling point elevations were measured with a simple ebullioscope and a differential thermometer.

Glover and Hill concluded that under the conditions of the above experiments both the solvent and solute affect the observed values of K_B. To avoid concentration effects, the work was carried out mainly at low concentrations (all less than 0·15 molal, usually about 0·05 molal). Their results showed that *in certain solvents* the experimental value of K_B increases as the molecular weight of the solute used in its determination increases. However, in some other solvents the experimental value of K_B

appeared to be independent of the solute molecular weight. Glover and Hill recommended that reliability could be improved by using solvents of the latter type in molecular weight determinations. It was noted that no variation of K_B with molecular weight was found using acetone and ethanol as solvent. In other solvents however there did appear to be a molecular weight dependence of the calibration constant which was not linear.

This probably indicates that the chemical nature of the solute as well as its molecular weight, influences the value of K_B. They therefore concluded that in relatively non-polar organic solvents (benzene and n-heptane) the experimentally determined value of K_B is dependent on molecular weight. In more polar solvents such as acetone and ethanol K_B is independent of molecular weight.

Glover stated that Barney and Pavelich[29] had reported the existence of such a dependence without giving any explanation. Similarly, Swietoslawski[30] had stated that K_B was dependent on molecular weight.

Glover[6] discussed this topic in some detail. It was noted that if the assumption is made that experimental measurements can be made on sufficiently dilute solutions or that corrections can be made for non-ideality, the experimental values of K_B (ideal) and K_B (experimental) should be independent of the solute as predicted by the relationship:

$$K_B = MWT_B/1000w$$

The present author is strongly of this opinion.

As Glover pointed out, a valid criticism by Lehrle[21] of the work of Glover and Hill[28] was that in the experimental study, no account was taken of the non-ideality of solutions. However this deficiency was remedied by a repeat of Glover and Hill's earlier work using a dual vessel ebullioscope, and evaluation of data to take into account non-ideality; however, a dependence of calibration constant on molecular weight was still observed. A divided difference calculation procedure was adopted and extreme curvature in earlier data was thereby eliminated. Nevertheless, Glover concluded that it did not appear from preliminary examination that elimination of superheating and measurement of the actual concentration of solution would eliminate the apparent dependence of K_B on solute molecular weight.

As a possible explanation Glover conjectured that, as a decrease in the vapour pressure of a solvent on addition of a solute is caused by a decrease in the number of solvent molecules at the solution surface, the relative effective size of the solute and solvent molecules may also be important.

Contrary to these observations are the findings of Daniels[24] in 1967 who did not observe a molecular weight dependence when using a small number of low molecular weight solutes including benzil. For instance, determinations of quite high molecular weights (e.g. 3000) produced experimentally determined calibration constants K_e for these solutes which gave results which agreed well with those by other methods.

Additionally Niwa and Noma,[31] in 1973, carried out molecular weight studies in Japan using a Takara model L-4 ebulliscope with benzene and acetone as solvents. They observed that the relation between $(\Delta div/c)$ where Δdiv denotes the change in experimental parameter, e.g. temperature or resistance, and c (concentration) gave a straight line relationship. However, the reproducibility of the experimental data was poor compared to that observed using vapour pressure osmometry.

More importantly they concluded that the values *for* K_e *for this particular ebullioscope were almost constant over a wide range of molecular weight*.

The authors concluded that best agreement between ebulliometry and vapour pressure osmometry was for a molecular weight of 2000–20 000.

7.3.   PRACTICAL EBULLIOSCOPY

7.3.1. General Aspects of the Technique

In view of the importance of the many practical aspects of this technique it is perhaps worthwhile to consider those which need special attention. This is made more difficult because to date the author is unaware of any ebullioscopes commercially available in the United Kingdom, which are suitable for the analysis of polymers.

Reference has already been made to the availability of a Takara Model L-4 ebullioscope[31] from Japan and to high precision ebulliographs EP68 and EP3, EP2 available in small-quantity production from SKB[16] (Institute of Organic Chemistry of the USSR Academy of Sciences).

Apart from these references other work in this field describes the use of a variety of purpose-built ebullioscopes. Consequently, description of a general operating technique is not appropriate; however a number of aspects of experimental technique do seem to be important and are therefore discussed. In general, though, the use of a standard technique appropriate to the type of equipment used will minimise the effects of variables in the system.

At the outset it should be noted that, as described,[24] the technique, although simple, is subject to five principal sources of error:

    (a)   superheating of the liquid in the ebullioscope,

    (b)   variation of the boiling point of the liquid caused by changes in atmospheric pressure,

    (c)   difficulty in determining the 'true' concentration of the liquid in the ebullioscope due to condensation of the vapour phase on the walls of the vessel—called the 'solvent hold-up',

    (d)   the existence of foaming resulting from the presence of polymeric solutes and the associated behaviour (known as sigmoid curvature) in experimental data,

    (e)   the presence of zero error observed in the plots of low concentration data points.

Many of the modifications in apparatus which have taken place have involved attempts to either reduce or eliminate these factors, particularly those concerning superheating and foaming. It has already been noted that the *superheating* of a boiling liquid may be related to several variables:

    (a)   the hydrostatic pressure at the temperature sensing point (i.e. below the liquid level) being greater than atmospheric pressure, and thereby affecting the pressure dependent boiling point of the liquid,

    (b)   raised vapour pressure due to the spherical nature of the bubble and,

    (c)   the requirement that the heater element and boiler surface need to be at a temperature greater than the boiling point of the liquid, for boiling to take place.

The earliest attempt to reduce the effect of superheating using a vapour lift pump, was described by Cottrell[4] and the effect estimated by Swietoslawski[30] to be of the order of 0·03 to 0·05 °C in certain cases. Consideration of the critical dimensions of the Cottrell type pumps indicates that although the dimensions of the arms do not appear critical, successful operation is not possible when the viscosity of the solution is appreciably greater than of that of an 'efficient solvent'.[5] By the term 'efficient solvent' is meant one that can be pumped easily.

Lehrle[21] has also referred to the encouragement of superheating if the surfaces of the boiler are smooth, most workers having used Swietoslawski's[30] method of overcoming this by sintering glass powder on to the sides of the boiler.

Other attempts to minimise the effect of superheating have included techniques for adjusting the power setting of the heater. Thus, Glover[12] has used a series of decreasing input values and the elevation as measured at the heating rate below which the solution was no longer insensitive to heat input. This procedure is particularly useful when dealing with polymer solutions which foam.

Finally, in respect of reducing the effect of superheating is the importance of effective insulation to protect the temperature sensors from fluctuation in ambient temperature. Such devices have, in the case of determinations of very high molecular weights, included the use of a vapour jacket which contains the same refluxing solvent as is used for the inner refluxing solution.[12]

More recent pumping systems have adopted mechanical means of pumping liquid in the ebullioscope, particularly that described by Daniels.[24]

There is still some uncertainty as to the exact cause of the problem of *zero error*. Nevertheless, it is probably related to a change in the surface behaviour of the boiling liquid caused by the presence of the solute which alters the hydrodynamics of boiling as suggested by Oliveira[32] or to a difference between the superheating of a boiling pure solvent and the superheating of a boiling solution.

More recent studies investigating the effect of zero error have been reported by Rapoport and Taits.[16] In an investigation of the factors affecting the accuracy of precision ebullioscopy, the authors observed that the interpretation of experimentally obtained ebulliograms was greatly affected by the tendency of the first point to give an erroneous value. It was further observed that this discrepancy was more marked in differential modes of calculation, whereas in integral calculations the effect was less serious.

Experimentally the authors noted that some workers had suggested preconditioning of the apparatus walls with the sample being used in order to reduce zero error.[19,33] Rapoport and Taits noted that from examination of their experimental results it appeared that only a part of the first aliquot of sample added to the solvent was effective in raising the boiling point of the solution. They adopted a graphical differential method to account for this defect and at the same time emphasised the importance of an atmospheric pressure correction in all calculations. A number of workers[4,15,21,29,34] have investigated the phenomenon of *foaming* in polymer solutions and its interference with the observation of the elevation of boiling point.

Lehrle[21] has discussed this effect in some detail emphasising that the viscosity of the polymer solution plays a large part in determining the lifetime of a foam. In addition Blackmore[34] and Smith[9] have concluded that foaming increases with the *breadth of molecular weight distribution* for a given molecular weight and that the dependence of foaming on concentration may vary with the magnitude of the concentration. A variety of approaches to overcoming the effect of *foaming* have been discussed. These include the possibility of a solvent change, adjustment of energy supply to the boiler heater, restriction of working concentration range and more recently, by alteration of ebullioscope design. A substantial investigation of the effect of foaming which can lead to the appearance of sigmoid behaviour in experimental results has been carried out.[26] Although several interpretations of sigmoid behaviour have been assessed none proved satisfactory. Nevertheless, the work showed that much of the sigmoid behaviour could be accounted for by variations in the degree to which superheating was dissipated with concentration which was most pronounced for viscous solutions.

The work of Washburn and Reid[35] involved early attempts to assess errors due to *solvent hold-up* by estimating the true concentration of solute in the liquid phase in a boiling solution.

The need arose because the concentration in the ebullioscope vessel boiler is different from that calculated on the basis of added solvent and solute. This discrepancy arises because part of the total solvent is removed from the liquid phase as condensing vapour. Glover[12] reported attempts to measure the distribution of concentration within the ebullioscope by removal of samples during its operation. However, results obtained at that time were still regarded as unsatisfactory. Other workers have concentrated on means of compensating for this error by adjustment of experimental conditions. In particular Hill and Brown[36] have made allowance for the effects of solvent hold-up by initially filling the ebullioscope with pure solvent. After draining and heating any excess solvent is deposited on the vessel's inner surfaces. The experiment then commences by delivering a known volume of solvent to the ebullioscope, the error due to solvent hold-up having been minimised. It has, however, been pointed out by many workers that an exact knowledge of the size of the solvent hold-up is not necessary provided a comparative method is employed. In this approach any error is compensated for by determining the experimental or apparatus ebullioscopic constant in a preliminary calibration using solutes of known molecular weight.

7.3.2. The Importance of Solubility

An extremely important aspect of the use of the technique of ebullioscopy is that it enables determinations of molecular weight to be carried out on samples whose solubility at ambient or near ambient temperature is very low.

Such a situation arises in the case of crystalline polymers which often show opalescence and dissolve easily only at temperatures near their melting points.

However, care must be exercised in correctly assessing whether solution has actually occurred. For instance, Blackmore[34] has reported that high molecular weight fractions in samples of polyethylenes do not dissolve in boiling carbon tetrachloride and the fact that solution has not occurred is not readily observable as the insoluble species float in the surface foam. Further, in the case of dark coloured samples it may not be readily apparent that total solution has not taken place. In this situation a procedure recommended[37] is that the regular response of the temperature to successive increments of sample should be monitored. A thorough study of the solubility characteristics of the sample in a variety of solvents should be carried out before ebullioscopic measurements are attempted.

To assist in determining the most suitable solvent on the grounds of acceptable solubility at the boiling point of the solvent, the concept of solubility parameters may be of some value.

For instance, in the case of amorphous polymers solubility is favoured where the heat of mixing is small. Lehrle[21] suggests that it should be possible to select a solvent whose solubility parameter (δ_1) is close to that of the solute (δ_2), whereby the heat of mixing ΔH_m is at a minimum. Values of δ_1 for a range of solvents are readily available from tables;[38] however, the concept outlined, although useful as a general guide, does not completely describe the criteria for solubility.

Finally, it is important to stress that although visual assessment of the solubility of a polymeric solute may be acceptable in most cases, there are other criteria to be considered. Most important amongst these is the need to establish that the sample is chemically unreactive with the solvent at its boiling point.

7.3.3. Solvent Choice and Purity

A number of authors including Glover[6,39] Lehrle,[21] and Bonnar et al.[37] have reviewed this subject in detail.

Although Glover[6] has suggested that the purity requirement for ebullioscopic solvents is often not as critical as is imagined, most workers in this field have taken particular care to ensure high purity, especially in respect of low water content. For instance, Daniels[40] used the purest available starting materials, viz. analytical grade reagent benzene and toluene and spectroscopically pure cyclohexane. These solvents were then dried with sodium wire and distilled using a fractionating column packed with Fenske helices. In this work, starting with a charge of 600 to 800 ml, the first 100 ml of distillate was discarded, the next 300–400 ml alone being collected and stored under nitrogen, over sodium wire and pre-activated molecular sieve. It was claimed that this rigorous procedure was responsible for greatly reducing ebullioscopic noise to a minimum.

There are, in addition, a range of requirements for solvents used in ebullioscopic experiments. For instance, care must be taken to ensure that such solvents are stable at their boiling points and do not undergo decomposition, dissociation, or association, particularly with the solute being studied.

Although, as will be discussed later (Section 7.5), azeotropic mixtures have successfully been used as ebullioscopic solvents, two phase azeotropes, particularly with water, must be avoided.

However, in this context Bonnar et al.[37] have proposed the use of water-solvent azeotropes as ebullioscopic solvents when it is not possible to dry completely an otherwise suitable solvent. Examples of such systems include, for instance, those water azeotropes involving ethanol and tetrahydrofuran.

With reference to the purity of ebullioscopic solvents it has been suggested[21] that low molecular weight impurities may affect the ebullioscopic experiment in other ways apart from those resulting from the presence of water. Thus, the presence of an involatile impurity, varying in amount in different batches of solvent will lead to inconsistency in performance of the solvent. In the opposite situation, claimed to be the more common, where the impurity is volatile, instability of temperature reading is observed.

Various means[41] of purifying solvents have been discussed and these sources should be referred to for further information.

Finally, mention should be made of two important factors in the choice of a suitable solvent for ebullioscopic measurements of molecular weight and these are discussed below.

First, of particular practical use, is the fact that solvent power for bringing about solution of the polymer usually increases with increasing temperature. In this respect the technique has a major advantage over other methods for determining the colligative property of molecular weight.

Secondly,[6] solvents having a higher ebullioscopic constant produce greater elevations of boiling point.[6]

7.3.4. Recommended Procedures and Precautions to be Taken

Even before an ebullioscopic experiment is carried out to determine number average molecular weight some consideration should be given to the treatment of a polymer sample, especially during the final stages of its preparation. The importance of this lies in the fact that the solvent used in the final stages of preparation or purification might well be the most suitable solvent for use in the ebullioscopic experiment. In this way errors due to the presence of what would otherwise be a low molecular weight impurity in the polymer would be avoided; freeze-drying a sample from solution may be better than precipitation of the sample at the end of the purification.

Regarding the steps taken to prepare the ebullioscope several views have been expressed. A common recommendation is that the ebullioscope should be well cleaned and dried before each determination. It has been stated that this step is unnecessary in routine operation, if the boiling temperature is recorded for each determination.[6] Nevertheless, the ebullioscope should be drained in the same way after each determination. It has been reported[10] that any traces of residual solute remaining after draining between determinations, may offset the effects of irreversible adsorption found on the surface of a particularly well cleaned ebullioscope.

Generally the choice lies between using an ebullioscope in the differential mode or apparatus in the form of twin ebullioscopes.

With respect to the addition of samples to the ebullioscope, these should be added so as to cause as little disturbance as possible to the liquid/vapour equilibrium.[6] At this point careful observation, especially where samples dissolve slowly, might indicate the existence of decomposition or even polymerisation. During the course of sample addition it is necessary to be quite certain that solution of sample has occurred before taking measurements of the elevation of boiling point.

7.4. THE INTERPRETATION OF EXPERIMENTAL DATA AND THE CALCULATION OF RESULTS

A number of authors have discussed aspects of this subject.[3,6,21] Having experimentally obtained boiling point elevation data, it becomes necessary to use these in conjunction with molar concentrations of solution in order to determine the molecular weight of the solute. If it becomes possible to assume that the solutions being examined behave as ideal solutions, then a value for the apparatus constant, K_A, may be obtained by use of the equation.

$$\theta = \frac{K_A}{M} 1000 \frac{W_2}{W_1} \tag{7.21}$$

or

$$K_A = \frac{\theta M W_1}{1000 W_2} \tag{7.22}$$

where θ is the magnitude of the measurement associated with elevation of boiling point, K_A is the apparatus constant, which incorporates the sensitivity factor relating the boiling point measurement to temperature inputs, and the solvent hold-up factor, W_2 is the mass of solute and W_1 the mass of solvent.

In this approach it is assumed that, in the case of ideal solutions, the elevation of boiling point is a function of the molal concentration.

However, most polymers in solution behave as non-ideal solutions and consequently particular emphasis must be placed on treatment of data arising from such experiments in such a way as to arrive at true values for molecular weight.

The presence of non-ideality implies that the elevation of boiling point is directly related not to molal concentration but to the activity of the solvent in solution. That we are essentially considering non-ideal solutions is made much more likely as at high molecular weight a dilute solution produces an elevation of boiling point which, in practice, may be too small to measure. Hence the need for working at higher concentrations—more likelihood of non-ideality—with extrapolation to infinite dilution.

We have already seen that a power series will represent the relationship of elevation of boiling point to concentration, thus

$$T = A_1^* C_0 + A_2^* C_0^2 + A_3^* C_0^3 \tag{7.23}$$

However, we also know that the possibility of zero error exists (i.e. the boiling point elevation–solute weight plot does not pass through zero on extrapolation) and the above expression may be modified to take account of this:

$$T = T_0 + A_1^* C_0 + A_2^* C_0^2 + A_3^* C_0^3 \tag{7.24}$$

where A_1^*, A_2^* and A_3^* are the first, second and third virial coefficients. Of immediate importance is the consequence that the first virial coefficient depends only upon the polymer molecular weight and the ideal ebullioscopic constant (i.e. $K_{B/M}$).

7.4.1. Calibration Experiment

Some aspects of the possible molecular weight dependence of the ebullioscopic constant have already been discussed (Section 7.2.3.). In the absence of a clear-cut opinion on this matter it is desirable to choose as calibration standard a compound which is as similar in chemical structure and molecular weight to the 'unknown' as is possible. For this purpose reference is made here to a range of molecular weight standards[42] supplied by a variety of manufacturers. It is recommended that three standards be chosen close in molecular weight to the unknown, ebullioscopic constants determined for each and a mean calculated.

7.4.2 Calculation Procedures

The two characteristics of particular importance in examining experimental data are:

(1) *Non-linearity* (Fig. 7.2). Generally observed at higher solute concentrations and indicates that the higher virial coefficients in the power series need to be taken into account.

(2) *Zero point error* (Fig. 7.3). Generally observable at lower solute concentrations and is quite common.

Zero error arises, it is thought, as a result of differences in the characteristics of boiling[21] caused by a change in the surface behaviour of a dilute solution. It is interesting practically to note that the first of a series of solute additions in the ebullioscopic experiment often produces rather more instability in the reading than do subsequent additions. Also, in the case of additions of polymer as solute, often a quite small initial addition will

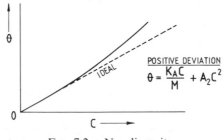

FIG. 7.2.   Non-linearity.

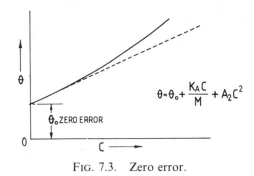

$$\theta = \theta_0 + \frac{K_A C}{M} + A_2 C^2$$

FIG. 7.3. Zero error.

produce quite noticeable foaming in the boiling solvent. In order to assess the possible presence of either non-linearity and/or zero error it is suggested that a plot of θ against weight concentration be made. However, in the case of polymer solutions it is probable that both errors will need to be taken into account. Two procedures are commonly employed to examine ebullioscopic data.

7.4.2.1. Limiting Slopes Plot
This technique uses the power series relationship previously described but neglecting the third term, thus:

$$\Delta T = A_1^* C_0 + A_2^* C_0^2 \qquad (7.25)$$

or

$$\frac{\Delta T}{W} = A_1^* + A_2^* C_0 \qquad (7.26)$$

Thus, the plot of $\Delta T/W$ against weight concentration is made and the intercept $W = 0$, gives A_1^* (A_2^*, the second virial coefficient may also be obtained as the gradient of the line).

The molecular weight of the solute can then be calculated from the relationship:

$$M = \frac{K_A}{A_1^*} \qquad (7.27)$$

where K_A is the apparatus constant and A_1^* the intercept.

Of considerable importance in the application of this calculation technique is the fact that it is likely to lead to erroneous results in the presence of zero error, which may not be readily detected unless it is

TABLE 7.1
Divided Difference Calculation—Generalised Approach

Solute additions	Weight of solute, W_i	Elevation, change in resistance, etc., θ	ΔW	$\Delta\theta$	$\Delta\theta/\Delta W$	$W_i + W_{i+1}$
—	0	0				
			$W_1 - 0 = W_A$	$\theta_1 - 0 = \theta_A$	$\dfrac{\theta_A}{W_A}$	$W_1 + 0 = W_A$
1st	W_1	θ_1				
			$W_2 - W_1 = W_B$	$\theta_2 - \theta_1 = \theta_B$	$\dfrac{\theta_B}{W_B}$	$W_1 + W_2 = W_B$
2nd	W_2	θ_2				
			$W_3 - W_2 = W_C$	$\theta_3 - \theta_2 = \theta_C$	$\dfrac{\theta_C}{W_C}$	$W_2 + W_3 = W_C$
3rd	W_3	θ_3				
			$W_4 - W_3 = W_D$	$\theta_4 - \theta_3 = \theta_D$	$\dfrac{\theta_D}{W_D}$	$W_3 + W_4 = W_D$
4th	W_4	θ_4				
			$W_5 - W_4 = W_E$	$\theta_5 - \theta_4 = \theta_E$	$\dfrac{\theta_E}{W_E}$	$W_4 + W_5 = W_E$
5th	W_5	θ_5				
			$W_6 - W_5 = W_F$	$\theta_6 - \theta_5 = \theta_F$	$\dfrac{\theta_F}{W_F}$	$W_5 + W_6 = W_F$
6th	W_6	θ_6				

sufficiently large to cause appreciable curvature in the plot of data. This calculation procedure also assumes that the third and higher terms in the power series are not significant. The necessity to use higher coefficients will be apparent if the plot shows curvature which increases with concentration.

By far the best approach and currently the most generally recommended method of treating data for polymer solutions is the method of divided difference described below.

7.4.2.2. Divided Difference

In this procedure zero error in data is examined in successive pairs of data points and not as in other methods, entirely at the initial point.

Thus, values are obtained by combining differences between successively lower values of θ and the sum and difference of successive values of W, the weight concentration.

The general approach describes results in the generalised relationship:

$$\frac{\Delta\theta}{\Delta C} = \frac{\theta_{(i+1)} - \theta_i}{C_{(i+1)} - C_i} = \frac{K_A}{M} + A_2(C_i + C_{(i+1)}) \qquad (7.28)$$

A plot of $\Delta\theta/\Delta C$ as ordinate is made against $(C_i + C_{(i+1)})$. If second-order non-ideality is present, the plot will have a gradient and should be linear with a value for K_A/M being obtained as the intercept to the slope.

The marked advantage of using this approach to examining the data is that if zero error is present (quite likely when using polymer solutions), the plot is unaffected.

The method of calculation for generalised values of solute additions (W_1, W_2, W_3, etc.) and associated elevations of boiling point (θ_1, θ_2, θ_3, etc.) is shown in Table 7.1.

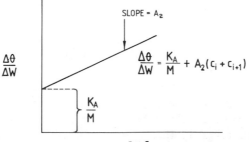

FIG. 7.4.   Divided difference graphical procedure.

A plot is then made of values for $\Delta\theta_A/\Delta W_B$, $\Delta\theta_B/\Delta W_B$, etc., as ordinate against W_A, W_B, etc., as shown in Fig. 7.4, and the intercept determined leading to a value of molecular weight in relation to the ebullioscopic constant, K_A. It is recognised that this calculation method will lead to a greater scatter in the data points obtained and the use of a linear regression technique is therefore strongly recommended.

7.5. RECENT DEVELOPMENTS

As already mentioned the major problem concerning the determination of molecular weight using polymer solutions is the experimental difficulty caused by foaming. Most attempts to reduce or overcome this problem have relied on modification to experimental procedure.

There are two aspects to the work:

(1) Modification of apparatus.
(2) Use of various solvent systems including azeotropes, applied pressure, etc.

Perhaps most *changes in experimental procedure* have involved modification of apparatus.

A number of workers have used rotation ebullioscopes to overcome the problems of foaming in polymer solutions. For instance, Sotobayashi et al.[17] have recently developed a new pressure controlled twin ebullioscope which does not utilise a Cottrell pump. A particular advantage of this system is the fact that the boiling temperature can be chosen by adjusting the pressure with a pressure regulator. The same authors noted that in their rotation ebullioscope foaming was eliminated when using samples of polystyrene up to molecular weight 4000 and other (unspecified) polymers. They concluded that the ideal requirement for vapour/liquid equilibrium was attained as demonstrated by the close agreement between the experimentally determined and the calculated theoretical ebullioscopic constants. In the experimental work described using polymer solutions with benzene as solvent it was shown that for polystyrene of molecular weight 4000 under atmospheric pressure, the plot, of boiling point elevation against molarity, passed through the origin. In the apparatus described it was possible to operate within a boiling temperature range of 30–80 °C.

In later work using another twin-type ebulliometer Yueh et al.[43] described the use of two boiling chambers for solution and solvent reference. It was claimed that low molecular weight materials could be determined using micro amounts of sample and that it was also possible to determine values greater than 10 000 molecular weight.

There have been very few recent references to the determination of very high molecular weights, i.e. up to 100 000. However, in 1974 Parrini and Vacanti[15] reported the use of a differential ebullioscope giving molecular weights up to 100 000 which were shown to have good reproducibility, accuracy and agreement with values obtained using other methods. In the work described, samples of unfractionated isotactic polypropylene were examined and it was shown that the method was applicable to polymer samples having low molecular weights or wide distributions.

Problems encountered using air or moisture sensitive compounds have been overcome to some extent by utilising a twin-type ebullioscope connected to a vacuum system.[44] Other attempts have been made to overcome apparatus deficiencies, such as the effect of solvent hold up, or dead space error.

When portions of solute are placed in an ebullioscope fitted with a drain tube, small pieces could become lodged in this part of the apparatus, perhaps remaining undissolved there. As a result incorrect values of molecular weight would be obtained.[45] To overcome this difficulty mercury was introduced into the drain tube and more accurate results obtained.

The only automatic system mentioned[46] consisted of an apparatus employing two Mariott bottles which held the samples and a series of valves. It was found possible to control the ratio of samples transported to the ebulliometer by adjusting the pressure on the Mariott bottles.

Another means[16] of overcoming the problem of solvent hold-up involved determining the molecular weight of compounds using a changing solvent volume. In essence it was found possible to determine the effective volume of solvent in the measuring cell. For the instrument used (a commercially available EP 2 precision ebulliograph (Special Design Office of the Institute of Organic Chemistry, USSR)), it was observed that the effective volume of the solvent was 0·83–0·99 ml less than the volume added to the cell. By taking into account this error substantially improved results were obtained. In practice the actual concentration of the solution in the ebulliometer was determined by making additions of a known concentration to the original solution.

Other ebullioscopic apparatus, notably that patented by Ambler,[47] has been claimed to have the advantages of both ebullioscopic and vapour pressure osmometer methods. In this technique there is provided a method for producing simultaneously separate continuous streams of solvent and solution, together with a saturated solvent vapour. Using a system of heat exchangers the two streams of solvent and solution are brought, as liquid films, into contact with the saturated solvent vapour and temperature differences are observed. The apparatus[47] utilises hypodermic syringe type pumps and thermistors as temperature detectors in a conventional Wheatstone bridge circuit. To ensure thermal insulation the envelope of the apparatus is surrounded by several vacuum jackets to maintain equilibrium conditions.

Of particular advantage is the fact that the equipment can be used up to 150 °C and at pressures above or below ambient. An unusual feature of the method is that either of the liquid streams may be changed from solution to pure solvent, or vice versa, simply by interchanging the connections. As a result a regular check can be kept on thermistor drift and bridge circuit balance.

Most of the work reported has involved the examination of solutes in organic solvent. A recent comprehensive discussion of the technique using aqueous solutions is described in work by De Oliveira.[32] The apparatus comprised a twin thermistor ebullioscope utilising a single-arm Cottrell pump. The diameter of the arm was relatively large (4 mm) with the tube of the pump being as long and wide as possible in order to minimise superheating effects. Thermal insulation was accomplished with poly-urethane foam, completely covering the ebullioscope and heating mantle.

Finally, the only recent reference to the use of a commercially available ebullioscope was that by Niwa and Noma[31] whose work involved studies using a Takara Model L-4 ebullioscope and an Hitachi–Perkin-Elmer model 115 vapour pressure osmometer with benzene and acetone being used as solvents. Samples of styrene polymers were examined using both techniques. Importantly, there was no evidence of a molecular weight dependence of the calibration constants in any of their work.

Although most methods to overcome the problems associated with the foaming of polymer solutions have involved apparatus modification as described, other approaches, particularly the choice of mixed solvent systems, have met with some success. Tsuchida et al.[48] have investigated the use of *azeotropes* in the ebullioscopy of polymers. They suggest that by this means it is possible to eliminate the effects of foam and its associated problems, viz. noise and drift. Using samples of polyethylene glycols they concluded that while azeotropic mixtures of cyclohexane/acetone, carbon tetrachloride/acetone, benzene/methanol, carbon tetrachloride/methanol and tetrahydrofuran/water were stable other azeotropes tested were less successful. A twin-type ebullioscope was used and all azeotropic mixtures were distilled twice with a helix packed column and then redistilled at atmospheric pressure immediately before use. The difficulty of taking in account errors due to solvent hold-up was overcome by calibration and the use of a twin system.

Further, it was found necessary to use a divided difference method of calculation in order to compensate for deviations observed in data at lower concentrations for both single solvents and azeotropes. The authors attributed this effect to negative zero point error and the adsorption of solute on the walls of the glass ebullioscope. The same apparatus has been used to determine the molecular weights of polystyrene in the 10 000 molecular weight region within a 1 % deviation.

In conclusion, it is seen that very little work with this technique is currently being reported. However, De Oliveira and Francisco[49] have described a simple ebullioscope which they used to determine boiling point elevations of $50 \times 10^{-6}\,°C$ which incorporated a single Cottrell pump transmitting boiling solution and vapour into a thermistor well. Modifications to the surface of the ebullioscope in contact with the heater mantle involved placing sintered ground glass inside the walls of the vessel. It therefore seems likely that future studies in the field of ebullioscopic determinations of molecular weight for polymers will be directed towards instrumental modifications.

These will presumably be aimed at coping more effectively with the two main problems associated with ebullioscopy, namely the removal of superheating during ebullition and the reduction of foaming in polymer solutions.

REFERENCES

1. J. H. van't Hoff, *Z. physikal. Chem.*, **1**, 481, 1887.
2. E. Beckman, *Physik. Chem. (Frankfurt)*, **4**, 532, 1889.
3. N. C. Billingham, *Molar Mass Measurements*, Kogan Page, London, 1977.
4. F. G. Cottrell, *J. Amer. Chem. Soc.*, **41**, 721 and 729, 1919.
5. R. Moore and B. M. Tidswell, *Chem. & Indus.*, **61**, 1967.
6. C. A. Glover, *Advan. in Analyt. Chem. & Instrumentation*, Interscience, **5**, 1966.
7. E. Zichy, *SCI Monograph No. 17*, 122, 1963.
8. N. H. Ray, *Trans. Faraday Soc.*, **48**, 809, 1952.
9. H. Smith, *Trans. Faraday Soc.*, **52**, 402 and 406, 1956.
10. C. A. Glover and R. R. Stanley, *Anal. Chem.*, **33**, 447, 1961.
11. C. A. Glover and J. E. Kirn, *Polymer Letters*, 3, 27–9, 1965.
12. C. A. Glover, *Polymer Molecular Weight Methods, Adv. in Chemistry Series*, American Chemical Society, M. Ezrin (Ed.), 1973.
13. G. R. Williamson, *J. Polym. Sci.*, A2, **5**, 1967.
14. K. G. Schon and G. V. Shulz, G. V., *Z. Phys. Chem.*, **2**, 197, 1954.

15. P. Parrini and M. S. Vacanti, *Die. Makromol. Chemie*, **175**, 935, 1974.
16. L. M. Rapoport and S. Z. Taits, *Russ. J. Phys. Chem.*, **49**(4), 518, 1975.
17. H. Sotobayashi, F. Asmussen and J. T. Chen, *Ber Bunsenges Phys. Chem.*, **81**(10), 1064, 1977; also *Makromol. Chem.*, **178**(11), 3025, 1977.
18. A. W. C. Menzies and S. L. Wright, *J. Amer. Chem. Soc.*, **43**, 2314, 1921.
19. W. Swietoslawski and J. R. Anderson, *Techniques of Organic Chemistry*, A. Weissberger (Ed.), Vol. 1, 3rd edn., Part I, Interscience, NY, 1959, Ch. VIII.
20. M. Ezrin, *Eastern Analytical Symposium on Molecular Weight Measurements*, New York, November 15, 1962.
21. R. S. Lehrle, *Progress in High Polymers*, Heywood, London, 1961.
22. W. R. Blackmore, *Rev. SCI. Instrum.*, **31**, 317, 1960.
23. R. S. Lehrle and T. G. Majury, *J. Polym. Sci.*, **29**, 219, 1958.
24. T. Daniels, *Ph.D. Thesis*, University of Birmingham, 1967; J. Foster, *M.Sc. Thesis*, University of Birmingham, 1964; J. Foster and R. S. Lehrle, *Europ. Polym. J.*, **1**, 117, 1965.
25. R. L. De Pablo, *Report to the University of Birmingham*, 1964.
26. T. Daniels and R. S. Lehrle, *Sigmoid Curvature in Ebulliometry—1*, *Europ. Polym. J.*, **6**, 1559, 1970.
27. W. G. McMillan and J. E. Mayer, *J. Chem. Phys.*, **13**, 276, 1945.
28. C. A. Glover and C. P. Hill, *Anal. Chem.*, **25**(9), 1379, 1953.
29. J. E. Barney and W. A. Pavelich, *Anal. Chem.*, **34**, 1625, 1962.
30. W. Swietoslawski, *Ebulliometric Measurements*, Rheinhold E., NY, 1945, p. 184.
31. M. Niwa and K. Noma, *Doshisha, Daigaku, Hokuku*, **14**(1), 60, 1973.
32. W. De Oliveira, *Differential Ebulliometry*, *Ph.D. Thesis*, Clarkson College of Technology, USA, 1975.
33. J. G. Van Pelt, *Philips Techn. Res. Reports*, **2**, 46, 1959.
34. W. R. Blackmore, *Canad. J. Chem.*, **37**, 1508 and 1517, 1959.
35. E. W. Washburn and J. W. Reid, *J. Amer. Chem. Soc.*, **41**, 729, 1919.
36. F. H. Hill and A. Brown, *Anal. Chem.*, **22**, 562, 1950.
37. R. U. Bonnar, M. Dimbat, and F. H. Stross, *Number-Average Molecular Weights*, Interscience, NY, 1958.
38. G. F. Price, *Techniques of Polymer Characterization*, P. W. Allen (Ed.), Butterworth, London, 1959, Ch. 1.
39. C. A. Glover, Polymer Molecular Weights, in *Techniques and Methods of Polymer Evaluations*, P. E. Slade (Ed.), Marcel Dekker, Inc., NY, Part I, 1975.
40. T. Daniels, *Ph.D. Thesis*, University of Birmingham, 1967.
41. J. A. Riddick and E. E. Toops, Organic Solvents, in *Techniques of Organic Chemistry*, Vol. VII, 2nd edn., Interscience, NY, 1959.
42. Polymer Laboratories Ltd, Shawbury, Shrewsbury, Shropshire.
43. H.-C. Yueh, S.-J. Wang and H.-Y. Hsu, *Hua. Hsueh Tung Pao*, **6**, 350, 1978.
44. B. J. Bulkin and P. J. Tergis, *Chem. Educ.*, **56**(4), 280, 1979.
45. K. Miyahara and T. Takaoka, *Bunseki Kagaku*, **23**(5), 515, 1974.
46. H. Koropecka and I. C. Koropecky, Czech. Patent 169, 923, 1977.
47. A. E. Ambler, British Patent 1,231,006, 5 May 1971.
48. E. Tsuchida, T. Kawata and I. Shinohara, *Makromol. Chemie*, **134**, 139, 1970.
49. W. De Oliveira and G. Francisco, *Chem. Biomed. Environ. Instrum.*, **10**(2), 189, 1980.

Chapter 8

ANALYSIS OF POLYMERS BY GEL PERMEATION CHROMATOGRAPHY

ANTHONY R. COOPER

Chemistry Department,
Lockheed Missiles and Space Company, Inc., California, USA

8.1. INTRODUCTION

Gel permeation chromatography (GPC) is a technique that separates molecules according to their size in solution. The principal use of the method has been for the characterisation of molecular weight distribution (MWD) of polymers, but the technique has also been used in the analysis of oligomers and low molecular weight compounds. The earliest use of GPC was for the analysis of water soluble polymers using cross-linked dextran gels[1] in the late 1950s. These soft gels were capable of excellent separations, but analysis times were long because the gels compressed at higher flow rates.

In the early 1960s a rigid gel was developed[2] for use in organic solvents. This extended GPC to the analysis of·commercially important synthetic polymers. These rigid gels also allowed the analysis to be performed much faster than could be done with soft gels. In recent years significant advances have been made in the development of additional rigid gels both for aqueous and non-aqueous systems. The introduction of high quality instrumentation, particularly sensitive, stable detectors and improved column technology has further reduced the analysis time.

This chapter reviews the current applications of GPC in both aqueous and non-aqueous systems. The primary topic considered is the determination of molecular weight distribution. Ancillary topics covered are the analysis of mixtures of polymers, branching, the compositional distribution of copolymers, and the purity of polymers. GPC has been the subject of several recent books,[3-12] book chapters,[13-24] and review articles.[25-40]

243

8.1.1. Theory of Gel Permeation Chromatography

The theory of the separation mechanism in GPC has received considerable attention. Many reviews include extensive discussions of the proposed theories of the separation mechanism (see refs. 14, 18–20, 23, 25, 27, 28, 30). Equation (8.1) is an empirical expression relating the GPC elution parameters:

$$V_R = V_0 + k_{GPC}V_i \qquad (8.1)$$

where V_R is the peak retention volume, V_0 is the interstitial volume, V_i is the volume of liquid within pores, and k_{GPC} is the distribution coefficient. Generally k_{GPC} varies between zero for excluded species and one for totally permeating species.

The mechanisms proposed to support the experimental observations (eqn. (8.1)) have been reviewed.[41] Equilibrium models propose that distribution between the 'mobile' phase and 'stationary' phase is a consequence of thermodynamic equilibrium, in which case the retention volume of any species would be independent of flow rate. The steric exclusion model postulates that the size of the molecule and the distribution of pore sizes determine the elution volume.

Another approach based on the equilibrium model is that k_{GPC} is a true equilibrium constant, with its associated free energy change on going from the 'mobile' phase to the 'stationary' phase. In the absence of an enthalpy change, the distribution is related to the entropy change. This has allowed Casassa[42–45] and Giddings et al.,[46] to calculate k_{GPC} for various model situations. It also explains separation by GPC with only a narrow pore size distribution. This had previously been explained on the basis of restricted diffusion theories which required that elution volumes be flow rate dependent. In the restricted diffusion theory[47–49] the rate at which species diffuse into the pores is slow in comparison to the residence time in the vicinity of the pore. Separation in this theory is controlled by the effects of molecular size on the diffusion kinetics.

Efforts to elucidate the exact mechanism of separation have been pursued and discussed.[50–54] Some recent results on the origin of flow rate dependence in GPC have suggested that a flow rate dependent equilibrium distribution coefficient is responsible.[55,56] The influence of polymer–sorbent interaction has been considered by several authors.[57–61] The effects which cause deviations from 'ideal' GPC include the cases where enthalpic interactions are present and may be negative, resulting in adsorption, or positive, resulting in partial exclusion due to incompatibility. The practical uses of adsorption chromatography of polymers has been reviewed[62] and are of significant use and complementary with GPC measurements.

8.1.2. Practice of Gel Permeation Chromatography

An ASTM method has been issued[63] describing the application of GPC analysis to polystyrene. Many manufacturers have chromatographs which are suitable for GPC operation at or near ambient temperature. The current requirements for GPC include a pump which is capable of stable flow rates in the range 0·1 to 3·0 ml min^{-1} and a high pressure injection valve, both capable of operation up to ~3000 psig. The column compartment should be temperature controlled, and detectors should have low volume flow cells (~5–15 μl) and be stable and sensitive. When polymers are insoluble at ambient temperature the requirements are more stringent, and all components of the flow system must be capable of operation at elevated temperature, in some cases as high as 130 °C. Only one commercially available instrument is able to operate at this temperature (Waters Associates Model 150C).

The columns for GPC are either packed with a porous cross-linked polymer or a porous inorganic material. These have been extensively reviewed.[14,18,19,22,31,32,37-40] Most often polystyrene–divinylbenzene gels are employed for non-aqueous GPC, but in certain applications porous glasses[64,65] or silicas[66,67] are preferred. The choice of column packing in aqueous GPC is more complex and is discussed in detail later in this chapter.

In the early literature, GPC columns are generally 4 ft long and 0·375 in outer diameter, and the packing particle size was 40 μm or above. More recently, column packings (termed micro-particulate) with improved efficiency have been produced by reducing the particle size to 5 μm or even less. Columns packed with these materials are generally about 2 ft long and 0·375 in outer diameter. Column packings have a discrete separation range so that a bank of columns is usually installed in the chromatograph to cover the molecular weight range of interest. The only recent development in this area has been the bimodal approach of Yau et al.[68] In this case the pore size of two columns is specifically matched to yield a linear calibration plot. Further attempts at reducing analysis times and solvent consumption have produced columns with only 0·5 mm bores.[69,70]

Some major advances have occurred recently in the area of concentration detectors suitable for chromatography. Most common is the use of differential refractive index and ultraviolet absorption, either fixed wavelength or tunable devices. Both are extremely useful for GPC but do however require calibration for their response factors at low molecules weights[71,72] to obtain quantitative concentrations of oligomers and additives. In ultraviolet detectors the major advance has been the introduction of diode array devices which obtain a full spectrum in ~ 1 s. This makes possible on-line identification of polymer or additive and is applicable to the determination of copolymer and multi-mer polymer distributions.[73,74]

The application of infrared detectors is increasing. Techniques have included single-wavelength monitoring,[75-77] scanning with stopped flow,[78] and computerised i.r. of collected fractions.[79] The applications of this detection mode are expected to increase with the increasing availability of Fourier transform infrared spectrometers.[80,81]

Detection of polymers in GPC effluents has been achieved by density measurements.[82-88] Linear response was found in the range of concentrations encountered in GPC effluents. Its requirements for thermostatting were similar to those needed for differential refractometers.

Determination of the mass of solute in GPC effluents has been attempted. One approach used a piezoelectric quartz mass sensor capable of detecting 10^{-10} g changes.[89] The GPC stream was sampled, and then the solvent was evaporated. Data points were obtained at 1 min intervals for butyl rubber and polystyrene. These data compared well with refractometric results.

An evaporative analyser has been described,[90] where the effluent is passed through an atomising head into a heated evaporative column. The solvent is evaporated and the solute is formed into small particles. These are subsequently detected by light scattering at $120°$ to the incident beam. The accuracy of the detector appears to equal that of the differential refractometer. Its use would be advantageous with mixed solvents or when low refractive index differences between polymer and solvent occurred.

Radiotracer techniques have also been used to detect the mass of polymer eluting from a GPC column. Typically ^{14}C labelling is monitored.[91]

The measurement of flow rate has been troublesome in GPC, where syphons are typically used. Many studies dealing with the effect of flow rate are obscured by the non-constant dump volume of some syphons at different flow rates. Failure to calibrate the syphon can lead to erroneous results. Alternate designs of syphons have been reported to minimise this effect.[92] An alternate method[93] essentially averages a large number of syphon counts to eliminate errors due to individual dump volumes.

The errors caused by flow rate variations in GPC when the pump is assumed to perform at constant flow rate which is not monitored have been described.[94] A method based on continuous weighing of the column effluent has been reported.[95] This is a highly precise way to monitor elution volume.

Data reduction in GPC is time consuming and has therefore been analysed by computer techniques since the inception of GPC.[96] Currently, several manufacturers have computerised data systems which perform data acquisition and reduction functions. Operator interaction is often needed or even required to set integration limits for the calculations. Molecular weight calculations are very sensitive to setting baselines and to the tails of the chromatograms. Computer manipulation of chromatograms[97] and statistical methods for comparing GPC curves[98] have been described.

Detailed evaluations of the origins of chromatogram broadening have been reported[28,54,99,100] and reviewed.[18] Experimental investigations of the effects of operating parameters for microparticulate columns have been performed.[101-107] Applications demonstrating the exceptional resolution of these microparticulate columns have included a resolution of the 600 MW polystyrene standard to its component oligomer peaks in approximately two hours (Fig. 8.1).[108] Using a 20 ft column series and 5 μm particles the polystyrene standards were found to have polydispersities as low as 1·0006 at 0·5 ml min^{-1},[109] two of the standards were shown to be bimodal. Microparticulate columns have been used over a wide flow rate range of 0·5 to 4·0 ml min^{-1}. The advantage of these columns is that they can achieve resolution similar to the macroparticulate columns in much shorter times. Also, when time is not the important parameter they can achieve resolution which macroparticulate columns are unable to develop under any realistic conditions. Some soft gels are capable of resolution equal to the microparticulate columns, but analysis times are slower by an order of magnitude.

8.1.3. Determination of Molecular Weight Distribution by GPC

A complete analysis of the data available from the GPC experiment includes a determination of the molecular weight averages of a polymer sample and its integral and differential molecular weight distributions. This section deals with recent developments in this area.

8.1.3.1. Calibration of GPC Columns

Calibration may be performed by use of a series of narrow MWD standards, one or more broad MWD polymers with known MWD or MW averages, or by the universal calibration approach. These methods have been recently reviewed.[110]

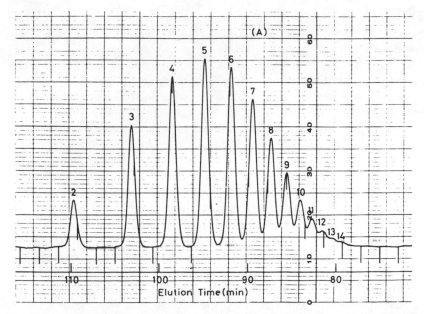

FIG. 8.1. GPC elution curve for polystyrene with molecular weight 600. The numbers refer to the degree of polymerisation. (Reproduced from Kato *et al.*[108] with permission of John Wiley and Sons.)

a. Use of narrow MWD standards

This method involves chromatographing several standards and plotting molecular weight versus elution volume. The number of polymers for which suitable standards are available is limited. After calibration the data points are treated mathematically so that the molecular weight at any elution volume is available either by interpolation, or polynomial curve fitting, etc.

Some thought has been given to the value of MW which should be plotted. Theoretical solutions were obtained[111] for standards which had log-normal, Schulz or Stockmayer distributions. The chromatogram peak was found to be $(\bar{M}_w . \bar{M}_n)^{0.5}$, $\bar{M}_w$ and $\bar{M}_n$, respectively. The viscosity average molecular weight $\bar{M}_v$ has also been proposed.[112] It was recently shown[113] that for either a log-normal or generalised exponential function that the first moment of the chromatogram is correlated with $(\bar{M}_w . \bar{M}_n)^{0.5}$ Errors arising from the use of linear calibrations or the incorrect polynomial expression (up to fourth order) have been determined.[114]

b. Calibration with broad MWD standards

This has been reviewed in detail.[110] Briefly, the polymer may have a known integral distribution[115,116] in which case the integral distribution curve is superimposed on the integrated normalised chromatogram. Then, known molecular weights may be correlated with known elution volumes. If the calibration curve is not linear, more complicated data fitting is required.[117] An empirical approach employing broad standards and a polystyrene calibration curve has been reported.[118] Many papers are available which describe iterative techniques to generate calibration curves from one or more broad distribution polymers. A simple method was developed for cases of linear calibration curves.[119] Several methods offering increased precision have been described.[120-125] Some comparisons of different methods have been reported.[126,127]

c. Universal calibration

Procedures have been extensively pursued as an alternative to calibration for each polymer type. Benoit et al.[128] proposed the product $M[\eta]$, where M is the molecular weight and $[\eta]$ is in the intrinsic viscosity, as the universal parameter. This procedure has been reviewed[129-131] and extended to rigid rods.[132] If the Mark–Houwink parameters (K and a from the equation $[\eta] = KM^a$) of a polymer are known, then under the assumption of universal calibration the curve determined for polystyrene may be used to generate the calibration curve for the other polymer. If K and a are not known, then they can be back calculated as described by Weiss and Cohn-Ginsberg[133] if two samples of that polymer are available with different intrinsic viscosities or one sample with known $[\eta] \bar{M}_n$. The reliability of data calculated from the universal calibration by the direct application of K and a has been compared with that of the Cohn-Ginsberg approach.[134] Recently it has been proposed that the universal calibration parameter should be expressed by $[\eta] . \bar{M}_n$.[135] An alternate approach, termed the southern method,[136] proposes that $\ln(V_R - V_0)$ plotted against $([\eta] . \bar{M}_w)^{1/3}$ gives a statistically superior universal calibration.

In generating the calibration curve attention must be given to the effects of polymer concentration. One effect of this is the decrease in hydrodynamic radius of the polymer with increasing polymer concentration, which changes the distribution coefficient and is observed as an increase in the elution volume. Viscosity effects in the interstitial volume cause directionally similar effects. Secondary exclusion which accounts for a reduction in accessible pore volume due to occupancy polymers causes a directionally opposite effect. The thermodynamic quality of the solvent may also be exploited to minimise or eliminate the dependence of hydrodynamic radius on solution concentration. Several theoretical treatments of the effects have been performed.[137-150] Interesting abnormal concentration effects have been described when polymers of different molecular weight are made a component of the solvent.[151] The presence of polymers in the solvent did not always increase the elution volume of the injected polymer as expected. In most cases concentration effects can be circumvented. In certain cases where detection is a problem or in preparative work high concentrations are unavoidable.

8.1.4. Molecular Weight Detection of GPC Solutes

Calibration in GPC can be simplified if the molecular weight of the eluting species can be determined. One approach has been to monitor the viscosity of the effluent in either a continuous manner[152] or by a series of measurements at discrete intervals.[153–159] The continuous viscometer detector is shown in Fig. 8.2. The pressure drop across a narrow capillary tube is monitored by a pressure transducer. The intrinsic viscosity is calculated at any time from eqn. (8.2):

$$[\eta] = [\ln (\Delta P_1/\Delta P_0)/C]_{C \to 0} \tag{8.2}$$

where ΔP_1 and ΔP_0 are the pressure drops due to the solution and pure solvent, respectively. The concentration, C, at any time is determined from

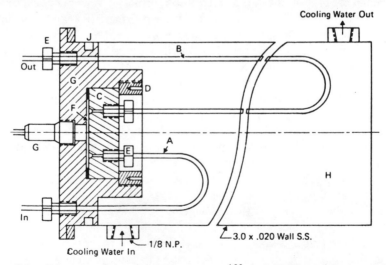

FIG. 8.2. Diagram of the viscometer system.[152] A, 0·020 in stainless steel inlet connection; B, capillary viscometer; C, pressure cell upper platen; D, upper platen retaining ring; E, swagelok fitting; F, pressure cell chamber; G, pressure transducer; H, thermostat; J, thermostat O-ring seal. (Reproduced from Ouano[152] with permission of John Wiley and Sons.)

the concentration detector response and the known weight of polymer injected. Typical pressure and concentration traces are shown in Fig. 8.3. If the universal calibration is valid, molecular weight values may be calculated even if K and a values are not available.

Similarly, the universal calibration curve may be constructed from samples with known molecular weight and unknown intrinsic viscosities. In the discontinuous measurements the approach is similar, but some

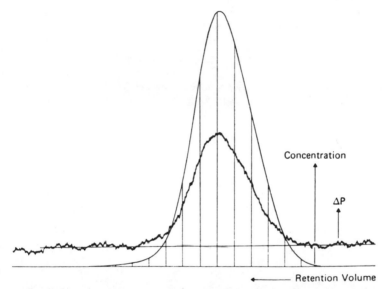

Concentration

ΔP

◄─── Retention Volume

FIG. 8.3.   Chromatogram of a narrow MWD sample showing concentration and ΔP traces. (Reproduced from Ouano[152] with permission of John Wiley and Sons.)

corrections are required to account for delay times and effluent hold-up in the syphon. Corrections for polydispersity of each fraction have also been applied. Theoretical evaluation of the errors caused by imperfect resolution of the GPC columns by continuous and discontinuous molecular weight detection has been reported.[160,161]

Continuous molecular weight determination may also be achieved by light-scattering techniques.[162,163] This is performed by measuring the scattering intensity at low angles from a laser source.

The molecular weight may be calculated from eqn. (8.3):

$$\frac{KC_i}{\bar{R}_{\theta i}} = \frac{1}{M_i} + A_i C_i \qquad (8.3)$$

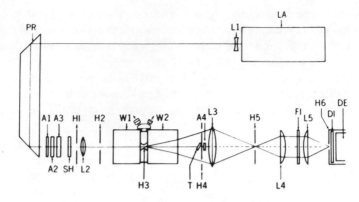

FIG. 8.4.   Schematic diagram of the optical system for the low angle laser light scattering detector. (Reproduced from Ouano and Kaye[162] with permission of John Wiley and Sons.)

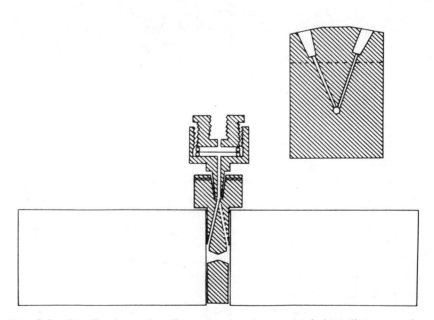

FIG. 8.5.   Details of sample cell construction for low angle laser light scattering detector. (Reproduced from Ouano and Kaye[162] with permission of John Wiley and Sons.)

where

$$K = \frac{2\pi^2 n^2}{\lambda^4 N}\left(\frac{dn^2}{dc}\right)(1 + \cos^2 \theta)$$

A_i is the second virial coefficient, C_i is the concentration and $\bar{R}_{\theta i}$ is the excess Rayleigh factor.

Thus, several parameters must be measured, some independently, e.g. dn/dc and A_i, and others continuously, e.g. C_i from a concentration detector and $R_{\theta i}$ from the calibrated recorder response.

The optical system is shown in Fig. 8.4 and the cell is shown in Fig. 8.5. Typical outputs are shown in Fig. 8.6. Further details of a commercially available instrument have been described[164] and its application has been

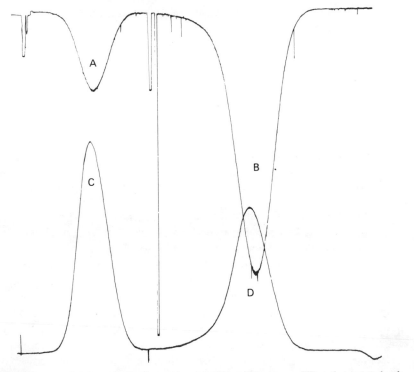

FIG. 8.6. Chromatograms of a mixture of polystyrene calibration standards having $\bar{M}_w$ $1\cdot8 \times 10^6$ and $5\cdot1 \times 10^4$ showing the concentration curves C and D and the scattered light curves A and B. (Reproduced from Ouano and Kaye[162] with permission of John Wiley and Sons.)

reviewed.[165] Comments similar to those for the viscometric detector concerning the effects of polydispersity of the sample in the measuring cell on the calculated results apply to this detector as well. Recently the detector has been operated at elevated temperatures to characterise polyethylene in 1,2,4-trichlorobenzene at 135 °C.[166]

8.1.5. Correction for Chromatogram Broadening

The observed chromatogram $F(V)$ is related to the true chromatogram $W(Y)$ by Tung's equation:[167]

$$F(V) = \int W(Y)G(V - Y)\,dy$$

where $G(V - Y)$ is a dispersion function. It describes the weight fraction of a solute that should elute at the retention volume Y but is dispersed and detected at retention volume V. Many mathematical procedures have been proposed to solve this equation to obtain $W(Y)$, and these have been the subject of a recent review.[168]

Further details have been discussed for correction of chromatograms obtained with mass and viscometric and light-scattering detectors.[169] The applications of an approximate approach (assuming a Gaussian broadening function) to the determination of polydispersity have recently been described.[170] An additional analytical solution of Tung's equation has recently been published.[171] Calibration for chromatogram broadening may be performed by measuring the differences between known and experimental molecular weight averages.[123,172,173] They may also be obtained from reverse flow measurements on polymers with uncharacterised distributions[174,175] by recycle techniques[176–178] or refractionation.[179]

8.1.6. Determination of Compositional Heterogeneity by GPC

GPC has been used to provide structural information on copolymers or mixtures of two homopolymers.[180] By use of two detectors each structural component may be monitored. Each additional component would require an additional detector. Although at one time this may have been thought impractical, detectors are now available that can rapidly obtain the spectrum of the detector cell contents. Fourier transform infrared[181] or diode array ultraviolet spectrophotometers[74,182] offer the possibility of characterising very complex structures or mixtures. Assigning a copolymer molecular weight at a given elution volume has been achieved[180] by interpolation on a log scale between the calibration curves for the component homopolymers using eqn. (8.4):

$$\log M_{\text{copolym}} = W_1 \log M_1 + W_2 \log M_2 \tag{8.4}$$

The validity of this assumption has been challenged[183] and the correct copolymer molecular weight should be calculated from combined GPC-intrinsic viscosity data which would require an on-line viscometer. This has recently been reported for a series of PVC copolymers.[184] The application of universal calibration to copolymers has been investigated.[154,185]

Block copolymers have been analysed by linear interpolation between homopolymer calibration curves,[186] and this method has been compared with the on-line measurement of viscosity. Little reason for any preference was found.[187]

Recently, lightly cross-linked microgels of styrene and divinylbenzene have been subjected to GPC[188] analysis and found to conform to the universal calibration if lithium bromide was added to the solvent (DMF).

Another approach to copolymer characterisation is to couple two GPC units operating with different solvents. In the first GPC unit a hydro-dynamic volume separation was achieved, and in the second GPC unit the separation was based on 'polarity'.[189,190]

8.1.7. Determination of Branching in Polymers

Branched polymers have also been characterised by GPC, and an extensive review has been made.[191] Analysis is usually based on the universal calibration approach, and it has been shown that they can be valid even for highly branched polymers.[192] The product of $[\eta].[M]$ is obtained for branched polymers from the universal calibration curve obtained using a homopolymer. The intrinsic viscosity for the branched polymer $[\eta]_{Br}$ is calculated from the intrinsic viscosity of the linear polymer $[\eta]_{Lin}$ from

$$[\eta]_{Br}/[\eta]_{Lin} = g^\chi$$

or

$$[\eta]_{Br}/[\eta]_{Lin} = h^3$$

where g is the ratio of the mean square radius of gyration of branched and linear polymers having the same molecular weight. Chi has been assigned the following values: $1/2$, $4/3$, $3/2$ and $(2-a)$, where a is the Mark–Houwink exponent.

These various branching functions have been compared,[193] and for comparative rather than absolute results the h^3 was found to be adequate at all branching densities. For the other functions $g^{1/2}$ was accurate for lightly branched polymers and $g^{3/2}$ was accurate for highly branched species. To circumvent the need to assign a parameter such as g^χ or h^3 an approach was used which involved the measurement of the intrinsic viscosity of the whole polymers. The measurement, combined with the universal calibration, yielded molecular weight distributions of the polymer and a parameter characterising the average branching frequency. Similar approaches have been used employing other methods such as sedimentation[194] or light scattering[195] to characterise the molecular weight of the whole polymer. The latter authors have critically reviewed the many variations of experimental approaches possible.

More recently the continuous monitoring of intrinsic viscosity or weight average molecular weight by light scattering has become available. Methods for determination of branching using a continuous viscometric detector has been developed.[196-199] Specific applications of this method to branching in polyethylene samples has been reported.[200-201] Light scattering detectors have been applied to determine branching for branched polystyrene[202] poly(vinyl acetate)[203] and polyethylene.[204]

8.1.8. Preparative Scale GPC

Relatively large amounts of polymer fractions with narrow MWD may be prepared by preparative GPC and used for fundamental physical

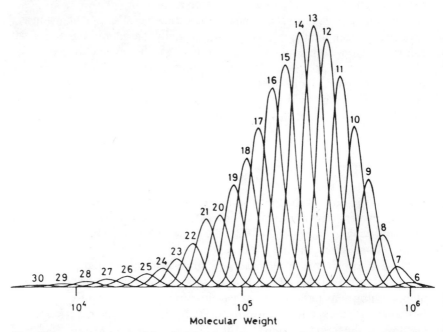

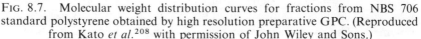

FIG. 8.7. Molecular weight distribution curves for fractions from NBS 706 standard polystyrene obtained by high resolution preparative GPC. (Reproduced from Kato et al.[208] with permission of John Wiley and Sons.)

measurements, mechanical property characterisations and for calibration purposes in the measurement of molecular weight. Since the columns are large and expensive and solvent volumes are high, system optimisation is very important. Such a study has been conducted[205] for columns of 2·25 in outer diameter. Reviews of preparative scale operation and applications have appeared.[206-207] The application of microparticulate packings in preparative GPC columns (2 ft × 0·83 in inner diameter) has been reported.[208] The effect of increased solute loading was compensated for using a θ solvent. Fractions of polystyrene having $\bar{M}_w/\bar{M}_n$ values of $\sim 1\cdot02$ were obtained (see Fig. 8.7 and Table 8.1).

Recent applications of preparative scale separations have included poly-(vinyl chloride),[209,210] poly(methyl methacrylate),[211] polypropylene,[212] poly(dimethyl siloxane),[213] and poly(acenaphthylene).[214] Modifications of the straight through technique to improve preparative scale operation have included recycle operation[215,216] and continuous operation.[217,218]

8.1.9. Calculation of Molecular Weight Averages and Distribution

When corrections to the chromatogram have been applied for band broadening, variation of detector response with composition or molecular weight, etc., a calibration curve is required for the polymer of interest. Then molecular weight information may be calculated. Molecular weight averages may be calculated from a series of chromatogram heights, h_i, measured at equal elution volume intervals. Molecular weights are assigned to those points from the correct calibration curve. The number, weight, and Z average molecular weights are calculated from eqns. (8.5), (8.6) and (8.7), respectively.

$$M_n = \frac{\sum h_i}{\sum (h_i/M_i)} \tag{8.5}$$

$$M_w = \frac{\sum h_i M_i}{\sum h_i} \tag{8.6}$$

$$\bar{M}_z = \frac{\sum h_i M_i^2}{\sum h_i M_i} \tag{8.7}$$

These averages should equal those determined by the classical methods of osmometry, light scattering and ultra-centrifugation.

More information is available from the GPC experiment which is a description of the integral and differential molecular weight distribution. The normalised chromatogram heights may be summed and plotted

TABLE 8.1

Details of Fractionation of NBS 706 Polystyrene by High Resolution Preparative GPC[208]

Fraction number	Weight, mg	$M_n \times 10^{-4}$	$M_w \times 10^{-4}$	M_w/M_n
6	2	109·	111·	1·017
7	8	90·8	92·5	1·019
8	20	76·7	78·0	1·018
9	42	62·3	63·5	1·020
10	64	51·5	52·5	1·020
11	84	41·7	42·5	1·020
12	102	34·7	35·4	1·020
13	110	28·7	29·3	1·022
14	110	23·3	23·8	1·022
15	96	18·9	19·4	1·022
16	86	15·9	16·3	1·023
17	68	13·0	13·3	1·022
18	54	11·0	11·2	1·022
19	42	9·14	9·35	1·022
20	30	7·52	7·68	1·021
21	28	6·16	6·28	1·020
22	18	5·12	5·22	1·020
23	12	4·09	4·18	1·021
24	8	3·33	3·40	1·021
25	6	2·63	2·69	1·024
26	5	2·01	2·06	1·026
27	4	1·52	1·56	1·028
28	3	1·13	1·16	1·031
29	2	0·83	0·85	1·033
30	1	0·59	0·61	1·035
Total	1005	13·7[a]	27·2[b]	1·99[c]

[a] $\sum w_i / \sum (w_i/M_{ni})$.

[b] $\sum (M_{wi}w_i)/\sum w_i$.

[c] b/a.

Reproduced from Kato et al.[208] with permission of John Wiley and Sons.

against molecular weight to give the integral distribution. This may then be numerically differentiated to give the differential molecular weight distribution. The differential distribution may be obtained from the chromatogram directly. If the calibration curve is linear, then the elution volume axis may be directly converted to molecular weight. If this is not the case then both ordinate and abscissa must be corrected.[219]

The differential distribution is a plot of dw/dm versus M, where w is the weight fraction with molecular weight below M. This may be written as:

$$\frac{dw}{dM} = \frac{dw}{dV_e}\frac{dV_e}{d(\log M)}\frac{d(\log M)}{dM} \tag{8.8}$$

$$= \frac{1}{M}\frac{dw}{dV_e}\frac{dV_e}{d(\log M)}$$

The ordinate of the chromatogram is dw/dV_e. The inverse of the calibration curve is $dV_e/d(\log M)$. If the calibration curve is linear, the latter term is constant, but if it is non-linear, numerical differentiation of the calibration curve must be performed at each point. Sometimes it is more convenient to plot the differential distribution in terms of $\log M$:

$$\frac{dw}{d(\log M)} = \frac{dw}{dV_e}\frac{dV_e}{d(\log M)} \tag{8.9}$$

Thus, in the case of a linear calibration curve the differential distribution plotted on a $\log M$ scale will be identical with the elution curve plotted on a retention volume scale.

The two ordinates dw/dM and $dw/d(\log M)$ are related by:

$$\frac{dw}{d(\log M)} = \frac{dw}{dM}\frac{1}{2\cdot303M} \tag{8.10}$$

From the differential distribution the molecular weight averages may be calculated from eqns. (8.11)–(8.13):

$$\bar{M}_n = 1 \bigg/ \int \frac{1}{M}\left(\frac{dw}{dM}\right)dM \tag{8.11}$$

$$\bar{M}_w = \int M\left(\frac{dw}{dM}\right)dM \tag{8.12}$$

$$\bar{M}_z = \frac{\int M^2\left(\frac{dw}{dM}\right)dM}{\int M\left(\frac{dw}{dM}\right)dM} \tag{8.13}$$

Similar expressions are available in terms of $dw/d(\log M)$. A computer program which performs these calculations[96] has been widely distributed. When the chromatogram contains discrete oligomer peaks or partially resolved peaks, then special methods of data reduction may be required.[220,221]

8.2. ORGANIC SOLVENT GEL PERMEATION CHROMATOGRAPHY APPLICATIONS

Since the commercial introduction of Styragel (cross-linked polystyrene gels) column packings and suitable instrumentation the applications of GPC have become widespread. It has become the single most important technique for the characterisation of soluble macromolecules. The introduction of other rigid packings based on porous glasses or silica and other cross-linked polymer packings has led to further improvements and applications for GPC. This section attempts to review, for a variety of polymeric materials, those combinations of column packing and solvent which have been shown to be applicable.

8.2.1. Polyamides
Polyamides have been chromatographed using m-cresol[222-224] or o-chlorophenol[225] and mixtures of chlorobenzene and m-cresol[226] with Styragel columns. Benzyl alcohol has been used for GPC of polyamides with silica gel packings.[227] These solvents require the use of high temperatures, typically above 100 °C, which may result in hydrolysis of the resin. They also have high viscosities which require the use of chromatographs that can operate at high pressures. Similar remarks apply to the use of dimethylacetamide.

Several new methods for characterising nylon resins have been proposed. One uses as a diluent chlorobenzene to reduce the viscosity of m-cresol[226] and allows the chromatograph to be operated at lower temperatures. Other solvents such as hexafluoroisopropanol used at room temperature[228] and 2,2,2, trifluoroethanol at 50 °C[229] allow characterisation of nylons to be performed with Styragel columns. Other column packings have been used, for example, deactivated Porasil and hexamethyldisilizane-treated porous glass have been used with a mixed solvent 50/50 m-cresol/chlorobenzene for GPC of nylons at 80 °C.[230] Yet another solution has been proposed; by N-trifluoroacetylation the nylon becomes soluble in common organic solvents[231] and also allows the use of the more sensitive u.v. detector. Methylene chloride solvent and Styragel columns have been used successfully in this approach.

Recently, aromatic polyamides have been prepared which are of considerable commercial interest. These materials have extremely restricted solubility. Solubility may be achieved in N,N-dimethylacetamide (only with the addition of a lithium salt in some cases) or concentrated sulphuric acid. A chromatographic method using 96 % sulphuric acid as solvent has been described[232] using Porasil-packed columns and u.v. detection at 330 nm. Similar studies using porous glass or silica with strong acids, methane sulphonic acid, 96 % sulphuric or 85 % phosphoric acid with a heterocyclic polymer.[233] Appreciable polymer absorbed to the column packing using 85 % phosphoric acid.

8.2.2. Polyesters

Aliphatic polyesters have been chromatographed[234] using tetrahydrofuran and Styragel columns at 37 °C. Tetrahydrofuran has also been used at 55 °C using Sephadex LH-20 and porous glass packings,[235] where a method for GPC calibration using vapour pressure osmometry results was used. The addition of acetic acid (1·0 to 0·1 %) to tetrahydrofuran has been recommended[236] for the analysis of esterification mixtures of terephthalic acid and ethylene glycol when using cross-linked styrene–divinylbenzene copolymers. Absorption effects due to the carboxyl group are suppressed.

The most common solvent used in GPC for poly(ethylene terephthalate) (PET) is m-cresol used in the range 110–135 °C.[237,238] The high temperatures required and the high solvent viscosity, together with the fact that PET can hydrolyse or undergo interchange reactions has led to the development of other chromatographic systems. Using Styragel columns, a solvent mixture of 0·5 % nitrobenzene in tetrachloroethane allowed room temperature operation of the GPC.[239] The use of hexafluoroisopropanol has been reported as a GPC solvent for PET[240] at room temperature with Styragel columns. A similar study using this system for the more difficult soluble poly(tetramethylene terephthalate) has been described.[241]

The analysis of PET oligomers has been performed by GPC after extraction from the polymer with chloroform.[242] Tetrahydrofuran was the GPC solvent employed with TSK gel-type H10 columns. A similar system was used to detect PET oligomers in refrigeration oils.[243] An investigation of the prepolymers of poly(propylene terephthalate) has reported the use of Bio-Beads S-X1 and -X2 with solvents such as chloroform, tetrahydrofuran, benzene and dioxane.[244]

8.2.3. Epoxies, Phenolics and Urea Resins

Epoxy resins have been characterised by GPC under various conditions depending on their molecular weight. Typically, tetrahydrofuran was used as the solvent and u.v. absorption as the detection mode. For low molecular weight resins, <1000, the use of Biorad SX-2 has been reported,[245] and combinations of Styragel and Biorad SX-2 for higher molecular weight materials. Merckogel OR 6000 has also been employed[246] to obtain high resolution chromatograms of low molecular weight resins. The use of an evaporative analyser has been reported[247] to quantitate oligomers which are difficult to isolate for calibration purposes.

GPC does not provide a total characterisation of these materials and is often used in conjunction with reversed-phase HPLC.[248–250] Poragel[248] and μ Styragel[249,250] column series have been used with tetrahydrofuran solvent for analysis of epoxy MWD and μ Styragel has been used with chloroform solvent to rapidly separate the resin from the curing agent.[249] Figure 8.8 shows a high-speed analysis of EPON 828 manufactured in

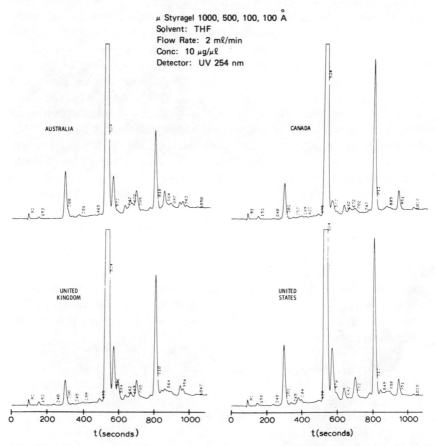

FIG. 8.8.   High resolution GPC of EPON 828. (Reproduced from Hagnauer[250] with permission of Department of the Army AMMRC.)

various countries.[250] Using a soft gel with high pore volume, epoxy oligomers up to DP 7 have been resolved.[251]

Determination of the molecular weight distribution and kinetic investigations of formaldehyde containing phenolics have been reported.[252–255] Typically Styragel columns were used with tetrahydrofuran solvent. A more complete analysis of these materials may be obtained by combining GPC results with reversed phase HPLC.[256]

More recently, Merckogel PVA 6000[257] has been used with tetrahydrofuran solvent to separate phenol or p-cresol–formaldehyde oligomers (Fig. 8.9). To determine the highest molecular weights formed in the reaction, Merckogel PVA 20 000 and 80 000 were used. Resorcinol–formaldehyde condensates have been characterised[258] using Bio-Beads S-X2 with tetrahydrofuran solvent. Columns packed with μ Styragel have been used with tetrahydrofuran or chloroform solvents to characterise the reaction products of formaldehyde with various model phenolic compounds to help understand the reactions of formaldehyde with condensed tannins.[259] The solutes were chromatographed as their acetates.

The use of N,N-dimethylformamide (DMF) as solvent with Poragel/Styragel columns allows analysis of phenol–formaldehyde, melamine–formaldehyde and phenol–formaldehyde–melamine condensates.[260] Improved resolution of methylolated phenols was obtained with this solvent. Similarly, the trimethylsilyl ethers of phenol and p-cresol and their methylolated derivatives have been separated using THF or ethyl acetate with Merckogel PVA 6000 columns.[261]

Urea–formaldehyde resins were initially analysed by GPC in water using Sephadex G or Bio-Gel columns.[262] Preliminary results using organic solvents were obtained[263] using Sephadex LH-20 and DMF solvent containing 2 % water. More rigid gels were tested, viz. Bio-Beads S-X2 and Merckogel[262] with pure DMF solvent. The inability of DMF to completely dissolve many urea–formaldehyde resins was overcome by dissolving the sample in dimethyl sulphoxide (DMSO). Once dissolved the sample did not precipitate during the analysis.

Another approach[264] to analyse urea–formaldehyde resins by GPC involved separation of low molecular weight materials in Sephadex LH-20 in water and the high molecular weight resin by using a Styragel–DMF combination. Reaction of alkylenediureas with formaldehyde forms oligomers with only even numbers of urea units.[265] This has been shown to improve the calibration of GPC systems with DMF and showed that Merckogel OR 6000 was able to separate higher oligomers than Sephadex LH-20.

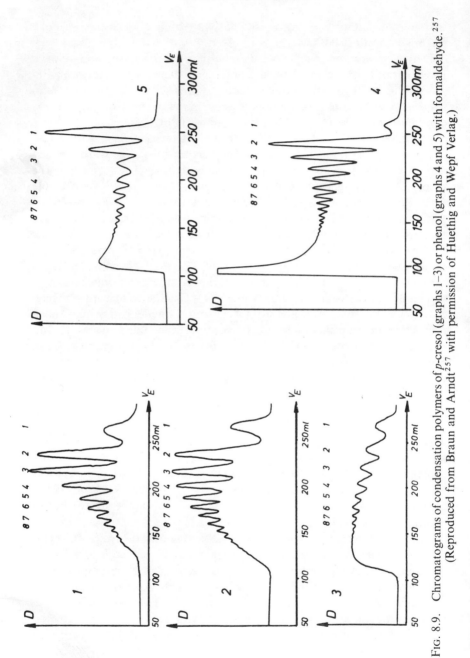

FIG. 8.9. Chromatograms of condensation polymers of *p*-cresol (graphs 1–3) or phenol (graphs 4 and 5) with formaldehyde.[257] (Reproduced from Braun and Arndt[257] with permission of Huethig and Wepf Verlag.)

Detailed analysis of the low molecular weight components is best approached by HPLC. The use of 85 wt % ethanol/water solvent with an anion-exchange resin and a cation-exchange resin were investigated.[266]

8.2.4. Polyurethanes

Polyester-based linear polyurethanes have been characterised[267] by GPC using Styragel columns and DMF solvent at 80 °C. In the absence of lithium bromide in the solvent multimodal chromatograms with apparent high molecular weight peaks were obtained. When lithium bromide was added to the DMF (0·05 M) the apparent high molecular weight peaks were absent (Fig. 8.10). The quantitation of toluene diisocyanate monomer in urethane adhesives has been described.[268] Poragel columns were used with THF solvent to separate monomer and prepolymer which had been reacted with ethanol. A more rapid HPLC method has also been developed.[269]

8.2.5. Elastomers

The use of GPC for the determination of MWD of elastomers is well established, typically using Styragel columns with THF solvent. The analysis of natural rubbers from 33 plants native to the north temperate zone has been reported.[270] The MWDs were found to be in the range used in commercial rubber production. A commercial high cis-polybutadiene fractionated by solvent gradient column chromatography was analysed by GPC and the fractions were shown to consist of two approximately log normal distributions.[271] A detailed study of the characterisation of polybutadiene (essentially 100 cis microstructure) with high molecular weight and broad MWD has been reported[272] with particular emphasis on the determination of number average molecular weight, M_n. The universal calibration approach has been investigated for ethylene–propylene copolymers and terpolymers[273] and found to be valid for the polymers studied.

Several investigations have used GPC to investigate the functionality distribution in elastomers. The olefinic unsaturation of several polymers containing low levels of unsaturation have been determined.[274] The method involved reacting the double bond with 2,4-dinitrobenzene-sulphenyl chloride. The use of dual detectors, RI (refractive index) and u.v., allowed the distribution of the original olefin groups to be established. A similar approach to determine hydroxyl functionality distribution in hydroxyl terminated polybutadienes has been used.[275,276]

The long chain branching in polychloroprene has been determined using GPC using the universal calibration curve, which linear polychloroprene was found to obey, and intrinsic viscosity measurements.[277]

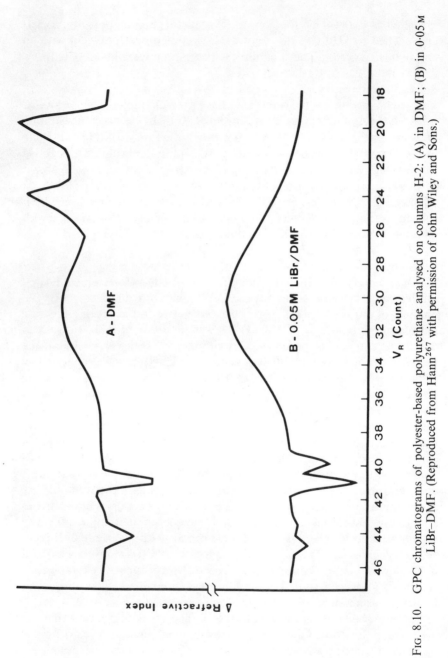

FIG. 8.10. GPC chromatograms of polyester-based polyurethane analysed on columns H-2: (A) in DMF; (B) in 0·05 M LiBr–DMF. (Reproduced from Hann[267] with permission of John Wiley and Sons.)

The results of GPC combined with thin layer chromatography have been reported[278] to evaluate the structure of polybutadiene and its styrene copolymers. On-line process analysis of the styrene–butadiene copolymerisation by GPC has been described.[279] Detection was achieved by a dielectric constant detector which is largely independent of styrene content (Fig. 8.11). Comonomer distribution in butadiene–styrene[280] and butadiene–α methylstyrene[281] have been studied by GPC using dual detection, RI and u.v.

The measurement of gel elastomers has been achieved by GPC. The gel content of nitrile elastomers has been determined[282] using methyl ethyl ketone as solvent with porous glass column packing with pore diameters of 677 Å and 75 Å. The differential refractometer signal was identical for soluble and insoluble species. A similar study using tetrahydrofuran for those elastomers which are not soluble in MEK has been described.[238] For styrene–butadiene copolymers a combination of 1400 Å and 80 Å porous glasses was used. Calibration of the differential refractometer was performed by bypassing the column. A portion of the material termed 'large gel' was retained by the column and calculated by difference.

8.2.6. Polystyrene
The MWD of polystyrene has been extensively studied by GPC which reflects in part that it was the first polymer to become available in the form of well characterised narrow MWD material. It is widely used to generate the universal calibration data for many GPC systems. The use of the larger particle size column has largely been superseded by the smaller particle size materials ≤ 10 μm. The use of μ Styragel,[284] 20 μm porous silica[285] and 5 μm-TSK gel[286] have been employed with tetrahydrofuran solvent. The results indicate that molecular weight averages obtained do not require corrections for instrumental broadening. The analysis time is reduced considerably over the larger particle size packings which can also produce results not requiring correction.[287,288] In these papers GPC was used to detect high or low molecular weight tails in supposedly unimodal polymer. It has also produced the most convincing evidence to date that NBS 705 and DOW S-109 polystyrenes are the same sample.

The importance of the role of solvent in GPC of polystyrene with styrene–divinylbenzene column packings has been investigated.[289–291] In chloroform solvent polystyrene, poly(dimethyl siloxane) and polyisoprene all conformed to the universal calibration plot. In cyclohexane, however, polystyrene is displaced to higher elution volumes in this θ solvent. The displacement increased as the elution volume increased and the

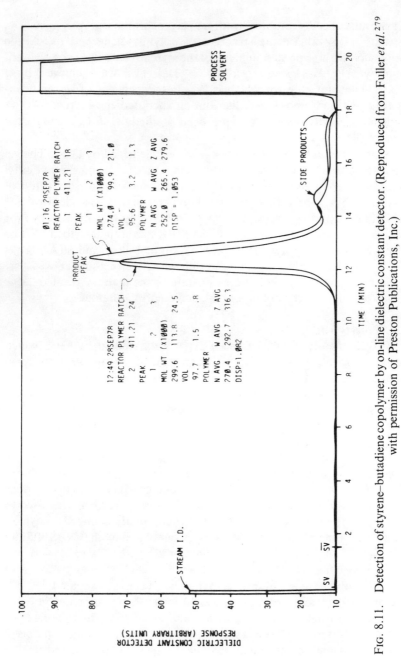

FIG. 8.11.   Detection of styrene–butadiene copolymer by on-line dielectric constant detector. (Reproduced from Fuller et al.[279] with permission of Preston Publications, Inc.)

chromatograms were broader than in chloroform and the broadening increased with increasing elution volume. Other θ or poor solvents for polystyrene, viz. *trans*-decalin and N,N-dimethylformamide, produced similar displacements to higher elution volumes. For *trans*-decalin at 100 °C the deviation from the universal calibration plot was small.

The effect of water content at 2·0 to 8·9 wt % in tetrahydrofuran has been investigated[292] and shown to produce shifts in elution volume, the shape of the solute peaks and a negative (vacancy) peak at the elution volume corresponding to injection of water.

The application of GPC to characterise very high molecular weight polystyrenes has been described[293] but degradation in the THF–Styragel GPC system was reported. To reduce the dependence of elution volume on injected concentration, θ solvents, either methyl ethyl ketone (MEK)[294] or MEK/methanol (89·3/10·7, v/v)[295] were employed with polystyrene gels. Results using porous silica materials have also been described using tetrahydrofuran solvent.[67] The most useful pore diameter was found to be 5000 Å for polystyrenes with molecular weights in the range 2×10^6–20×10^6.

GPC has found many applications in investigating the degradation of polystyrene. The ultrasonic and peroxide initiated degradation has been reported,[296] also photo-oxidative decomposition results.[297]

Copolymers of styrene have been analysed by GPC. The azeotropic composition polymer with acrylonitrile (27·8 wt %) was investigated in DMF and DMF–LiBr–H_2O.[298] With increasing addition of lithium bromide the polystyrene calibration curve shifted to higher elution volumes to a point where MWDs could not be obtained. The copolymer when analysed in pure DMF was bimodal, but unimodal in DMF–LiBr–H_2O. Since the viscosity of the polymer was identical in both solvents the conclusion was reached that association was not responsible for the high MW peak in the DMF solvent. Rather, it was caused by an artefact at the void volume which disappeared in DMF–LiBr–H_2O because the polymer eluted later.

GPC has been used in conjunction with pyrolysis–gas chromatography to investigate the composition of acrylonitrile–styrene copolymers.[299] The dual detector method was also employed. For these studies only analytical scale fractionations were required and chloroform was used as the solvent with polystyrene gel (Shodex A 80 M) columns. Typical results are shown in Fig. 8.12 with good agreement between the two methods.

The analysis of lightly sulphonated polystyrenes (2–5 mol) has been attempted by GPC.[300] Several solvent mixtures were evaluated THF,

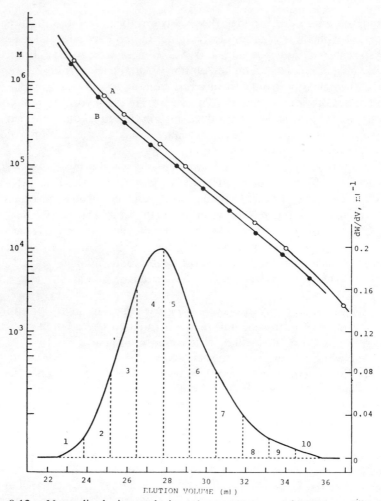

FIG. 8.12. Normalised size exclusion chromatogram, polystyrene calibration curve (A) and copolymer calibration curve (B) for styrene–acrylonitrile copolymer. (Reproduced from Mori[299] with permission of Elsevier Scientific Publishing Co.)

THF + 2 g litre^{-1} lithium nitrate, 10/1 THF/DMF and 10/1 THF/DMF + 2 g litre^{-1} lithium nitrate. Five μ Styragel columns were used: 10^6, 10^5, 10^4, 10^3, and 500 Å.

The polystyrene calibration curves were identical for all solvents except THF/DMF which showed late elution. For ammonium salts of the copolymer THF–LiNO$_3$ was found to be an effective solvent, for the sodium salt THF–DMF–LiNO$_3$ was required. In poor solvents multiple peaks were observed which were caused by adsorption.

8.2.7. Acrylics

Poly(methyl methacrylate) (PMMA) has been characterised extensively by GPC. It has been used as a standard for calibration of GPC systems where polystyrene is not soluble, e.g., in trifluoroethanol[301] with polystyrene gels. Fractions were produced by preparative GPC.

The universal calibration was found to be valid for PMMA, poly-(oxytrimethylene), poly(vinyl acetate) and polyamides in the trifluoro-ethanol–polystyrene gel system. Another study used preparative GPC[302] to fractionate PMMA in THF–Styragel system. Analytical GPC was then studied in n-butyl chloride (a θ solvent), THF, methyl ethyl ketone (MEK) and isopentylacetate with Styragel columns. Adsorption was observed in n-butyl chloride. The universal calibration for PMMA and polystyrene was observed in the THF but not in MEK or isopentylacetate.

The application of GPC to determine electron beam sensitivity of PMMA has been described[303,304] and allowed the effects of molecular weight and its distribution and tacticity to be evaluated. Poly(2-methoxyethyl methacrylate) has been fractionated in THF using porous silica.[305] The MWD of PMMA in acrylic bone cements has been determined with a DMF–Styragel combination.[306]

8.2.8. Polyolefins

One of the major triumphs of GPC has been the characterisation of polyolefins which are extremely difficult to characterise by other methods because of their limited solubility at room temperature. Early results were obtained with Styragel columns and 1,2,4-trichlorobenzene (TCB) solvent operating at 120–140°C. Calibration was performed usually with the narrow MWD polystyrene standards. Application to linear IUPAC polyethylene has been reported.[306a] Since column lifetime was a problem, alternative packings such as porous glass and silica were investigated. For porous glass the polystyrene standards would not elute due to adsorption. This was solved by treating the glass with hexamethyldisilazane which reacted with the surface hydroxyl groups.[77] These treated column packings were found to be extremely stable.

The application of GPC to determine branching in polymers,[307] and long chain branching in low density polyethylene[308] have been described and reviewed.[191] More recently, the use of a continuous recording viscometer has been used to investigate branching.[200,201]

The morphology of polyethylene has been investigated by etching with a suitable material which destroys the amorphous phase and the chain folds at crystallite surfaces.[309,310] GPC is then used to determine the number of ethylene units which traversed the crystallite.

Polypropylene has been characterised by GPC in systems similar to those used for polyethylene. Porous silica (Porasil) has been used[311] with TCB at 135 °C to analyse polypropylene. The results for MWD were in good agreement with column fractionation results. For the system Styragel/o-dichlorobenzene at 135 °C, the universal calibration has been found to be valid for polypropylene and polystyrene.[312]

Copolymers of ethylene and propylene have been analysed by GPC in a system of polystyrene gel and o-dichlorobenzene at 135 °C.[313] Copolymers of ethylene and vinyl acetate (~ 30 wt% acetate) have been analysed by GPC using a μ Styragel–THF combination at 20 °C.[314]

8.2.9. Vinyl Polymers
8.2.9.1. Poly(vinyl chloride)—PVC
Although PVC appears to be readily soluble in THF the polymer is not molecularly dispersed. Aggregates are formed increasingly with increases in syndiotacticity. In most cases where THF has been used as solvent with Styragel columns[315,316] or Styragel/Bio-Glass/Corning porous glass (CPG) combinations[317] at 25 °C some pretreatment of the sample solution was necessary to destroy aggregates. Temperature/time treatments of 200 °C/2 h^{-1},[317] 90 °C/90 min^{-1},[315] and 120 °C/3 h^{-1} (ref. 316) have been suggested as adequate for this purpose. Figure 8.13 demonstrates this phenomenon.[315]

Other approaches have been to use o-dichlorobenzene with 0·05 wt% Santonox R as anti-oxidant at 135 °C with Styragel columns.[318] A similar system was employed to preparatively fractionate a commercial PVC. No evidence of aggregate formation was found in this system. The use of N-methylpyrrolidone as an attractive solvent for GPC of PVC has been exploited.[319] Preparative fractionation of PVC has also been accomplished using silica beads (Porasil)[320] and Styragel.[321] In the latter it was found that chain branching caused higher $\bar{M}_w/\bar{M}_n$ values determined by light scattering and osmometry than by $(\bar{M}_w/\bar{M}_n)_{GPC}$. It was also found that the universal calibration curve for PVC and polystyrene was only valid at lower molecular weights where branching was negligible.

The use of GPC to analyse PVC for low molecular weight additives[322] and oligomers[323] has been reported. The former used Sephadex LH-60 with THF solvent and was useful for analysing a variety of additives, except

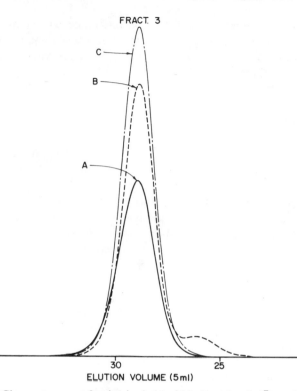

FIG. 8.13.   Chromatogram of poly(vinyl chloride). Fraction 3, $\bar{M}_w = 48\,700$: (——)
A, GPC effluent; (—·—·—) B, THF-soluble portion after precipitation; (———) C,
THF-soluble portion after heating in THF at 90 °C. (Reproduced from Salovey and
Gebauer[315] with permission of John Wiley and Sons.)

BHT and organotin stabilisers, without prior extraction. For oligomer
determination the polymer was pre-extracted with diethylether. Columns
employed were LH-60 Bio-Beads S-X3 and a column combination of μ
Styragel and PL gel.

GPC has been used to demonstrate the simultaneous chain scission and
cross-linking of PVC upon exposure to ionising radiation.[324]

The composition and molecular weight distribution of a poly(vinyl
chloride–vinyl acetate) copolymer has been determined[325] by GPC with
infrared analysis of films made from the fractions. No suitable solvent
could be found for on-line measurement. A system of polystyrene gel
columns (Shodex A 80M) with THF was used for the fractionation. A
mathematical treatment was described to calculate molecular weight

averages which included correction for the difference in refractive index as a function of composition and the difference in hydrodynamic volume of copolymers with compositional differences. This copolymer has also been characterised by preparative and analytical GPC using porous silica and Styragel columns.[326]

8.2.9.2. Poly(vinyl acetate)

The validity of the universal calibration approach has been examined for linear poly(vinyl acetate).[327] Using preparative GPC (Styragel–toluene) narrow MWD fractions of polymer were produced. The universal calibration was found to be valid in a Styragel–THF system at 35 °C. A similar approach was used for preparative GPC fractionation of branched poly(vinyl acetate).[328,329] The analytical GPC measurements were supplemented with viscosity and ultracentrifuge measurements. Application of GPC to study the shear degradation of poly(vinyl acetate)[330] has been reported.

8.2.9.3. Poly(vinyl alcohol)

An early paper[331] discussing the characterisation of poly(vinyl alcohol) employed a proprietary deactivated Porasil with water as solvent at 65 °C. More recently, reference samples have been produced by saponification of poly(vinyl acetate).[332] These samples were characterised by two GPC approaches. One involved the use of Spheron, poly(2-hydroxyethyl methacrylate); the other used TSK Gel PW, both with water as eluent.

8.2.9.4. Other Vinyl Polymers

Vinylidene chloride homopolymer is insoluble in THF but several copolymers have been analysed by GPC using a THF–Porasil system.[333] Comonomers studied included vinyl chloride, methyl acrylate, and methyl methacrylate.

8.2.10. Silicones

The preparation of narrow MWD poly(dimethyl siloxanes) (PDMS) and their characterisation by GPC has been reported.[334] The GPC method used a Styragel–toluene combination at 30 °C. Preparative GPC has been used to produce narrow MWD fractions of cyclic PDMS, again using a Styragel–toluene system.[335] The validity of the universal calibration approach for PDMS in a Styragel–toluene system at 60 °C has recently been demonstrated.[336] Copolymers of PDMS with polyether alcohols have been characterised[337] by using a Styragel–ethyl acetate system at 30 °C.

8.2.11. Cellulosics

Traditionally cellulose has been analysed by GPC, through a suitable, derivatisation to render it soluble in THF, using cross-linked polystyrene gels. For example, nitro, acetyl or carbanil derivatives have been employed and these methods have been reviewed.[338] The carboxymethyl derivative has been analysed by aqueous GPC.[339] Techniques not involving derivatisation have been used, notably the use of cadoxen (5–7 % cadmium oxide in 28 % aqueous ethylenediamine) as solvent. A novel modification of this uses cadoxen to dissolve the polymer but 0.5 N sodium hydroxide is used as the GPC solvent.[340] In this method Sepharose CL-6B was the column packing material. Inorganic supports have also been investigated[341] with cadoxen solvent.

A detailed study of the GPC of cellulose trinitrate[342] showed that ethyl acetate was a more appropriate solvent than THF. Also, in the former solvent the universal calibration was found to be valid.

The MWDs of hydroxypropyl methyl cellulose in tablet coating have been determined using a dimethylacetamide–Styragel system.[343] The analysis of paper coating starches has been performed with systems using either Bondagel or Syn Chropak columns.[344] Two solvent systems were used. One consisted of equal volumes of water and DMSO, the other was three parts water, one part DMSO (by volume), plus 10 mM phosphate. The oxidation (permanganate) products of lignin have been analysed using a DMF–Sephadex LH-20 system.[345] The use of DMSO with μ Styragel columns has been reported[346] for the analysis of polymers from wood.

8.2.12. Polar Polymers

Many polar polymers are soluble in water but have been analysed by GPC in an organic solvent because an aqueous GPC method was unavailable, or to reduce the number of times the instrument is subjected to a solvent changeover. This has not always proved to be a wise approach and several difficulties have been found when polar polymers are analysed in polar solvents.

Polyacrylonitrile and its copolymers with sulphonate containing vinyl monomers have been examined by GPC using a DMF–Styragel combination.[347] The homopolymer exhibited unimodal chromatograms whereas the copolymers showed multiple peaks. Anticipation that electrostatic effects were responsible for the multiple peaks led to the addition of 0.1 M lithium bromide to the DMF. The copolymers exhibited unimodal chromatograms in this solvent with later elution. The corresponding reduction in molecular size was also evident from viscosity measurements. A similar study of DMF with added salt (sodium nitrate) has been reported with Porasil columns.[348] It was found that 5×10^{-2} M salt concentration was required to suppress electrostatic exclusion.

Polyacrylamide and its partially hydrolysed derivatives have been characterised by GPC using porous glass packings with formamide as solvent.[349] For polyacrylamide the elution was delayed somewhat if potassium chloride was added (5×10^{-3} M). The universal calibration was observed for these systems. For the hydrolysed samples the elution was similar to the unhydrolysed sample in formamide. However, in the presence of potassium chloride the hydrolysed samples were delayed in elution by an amount far greater than could be accounted for from viscosity changes. Mixtures of polyacrylamide and its hydrolysis products showed two peaks when eluted with formamide–potassium chloride (5×10^{-3} M). Technical polyacrylamides also showed this elution behaviour.

The application of GPC to the synthetic polyelectrolytes has been reported[350] using DMF or dimethylacetamide (DMA) with porous glass and Styragel packings.

The universal calibration approach has been evaluated for the systems polystyrene, poly(2-vinyl pyridine) and their block or graft copolymers in THF and DMF[351] with Styragel columns. None of the systems (except polystyrene/THF–Styragel) conformed to the universal calibration.

The use of silanised glass columns has been suggested[352] for GPC of polar polymers in DMF to reduce adsorption and partition effects. In this system the effect of lithium bromide on polystyrene elution volume is less than 1 %.[353] For poly(N-vinylacetamide) the chromatograms showed later elution or bimodal behaviour in pure DMF compared with 0·01 M lithium bromide/DMF. This behaviour was ascribed to the formation of stable aggregates. Such molecular association has also been found in other systems such as DMF–octylsepharose CL-4B for the analysis of lignin.[354] Polymers which exhibited bimodal peaks in pure DMF were found to produce unimodal peaks when lithium chloride was added.

Aliphatic and aromatic polyacids and their derivatives have been characterised[355] by GPC using Styragel or porous glasses with DMF or DMA. In the absence of added electrolyte asymmetric peaks were obtained and their maxima did not correlate with molecular weight.

8.2.13. Miscellaneous Polymers

Polyisobutylene has been characterised in a variety of GPC systems. Agreement with results from gradient elution fraction atom has been demonstrated.[356] However, if the molecular weight is sufficiently high then degradation may occur.[357] These authors showed a dramatic decrease in M_w could occur in the GPC experiment with a polymer of initial $M_w = 1 \cdot 06 \times 10^6$. Laser light scattering detection was exploited to prove molecular weight reduction by mechanical degradation. GPC has also been employed to evaluate degradation during cold milling of polyiso-butylene.[358] Oil additive polymers of the polyisobutylene–succinimide type have been characterised by GPC[359] using μ Styragel–THF at ambient temperature.

Polycarbonates may be chromatographed in Styragel–THF systems[360] and the GPC results have been combined with light scattering and viscometry to elucidate branching parameters. The application of the universal calibration curve for this system has been shown to deviate at low MW.[361] Recently this approach has been improved by direct calibration with oligomers in the low MW region.[362]

Polysulphones have been characterised using Styragel columns using THF for poly(butene sulphone)[363] and DMF for poly(arylene ether sulphones).[364]

Alkyd resins have been characterised using Styragel columns with THF.[365] Multimodal peaks were obtained which were considered by these authors to be experimental artefacts, however this phenomenon would appear to warrant further investigation. The application of GPC to alkyd resins has been reviewed.[366,367]

Polyoxymethylene was characterised by GPC for the first time[368] using a Styragel–DMF combination at 140 °C, preparative scale GPC fraction was also performed.

Medically important polymers such as polyurethanes, silicone adhesives, and silicone moulding resins have been characterised with Shodex columns (Perkin-Elmer) and tetrahydrofuran solvent. Polyurethanes were detected at 247 nm, silicones were monitored by a highly sensitive RI detector.[369]

Fluoroether polymers have been characterised[370] by GPC in a system employing porous silica columns and Freon 113 solvent.

MWD analysis of Amylose[371] and lignin[372] have been reported using dimethylsulphoxide (DMSO) with deactivated porous silica and Sephadex G100, respectively.

8.3. AQUEOUS GEL PERMEATION CHROMATOGRAPHY APPLICATIONS

For many years aqueous GPC was limited to the use of a few commercially available organic gel column packings. These gels were of the soft variety and hence were applicable only at low pressures and low flow rates. The use of these types of materials has been described.[373–376] Their use was mainly in the field of biochemistry. Detailed reviews of applications of aqueous GPC to proteins[377,378] and polyelectrolytes[379] have been published. Reviews of aqueous GPC theory and applications have appeared.[380,381]

For neutral water soluble polymers, analysis by aqueous GPC is very similar to organic solvent GPC. However, for charged polymers there are some additional factors which need to be considered. The first effect is that of polyelectrolyte expansion where the polymer coil expands significantly as the ionic strength decreases causing elution at the column void volume. Sufficient electrolyte must therefore be maintained in the solvent to avoid this excessive expansion and be able to use the existing column packings. Another effect caused by charges on the solute is that of ion inclusion. This effect is depicted in Fig. 8.14.[382] A Donnan effect arises from the inability of

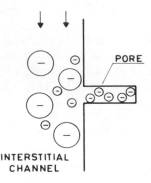

FIG. 8.14.   Ion inclusion of a low MW polyelectrolyte component into a pore in a gel. (Reproduced from Forss and Stelund[382] with permission of John Wiley and Sons.)

the largest charged species to enter the pore. This results in a higher concentration of smaller charged molecules in the pore volume than in the interstitial volume. The elution of the lower molecular weight species is delayed by the presence of the higher molecular weight polyelectrolyte. A further complication arises with electrolyte solutes if the column packing is charged. In this case, the sample, either a simple electrolyte[383] or a polyelectrolyte[382] exhibits badly skewed peaks (see Fig. 8.15), and this has been termed ion exclusion. The peaks always begin at the void volume and the shape is highly dependent on the amount of sample injected. At low

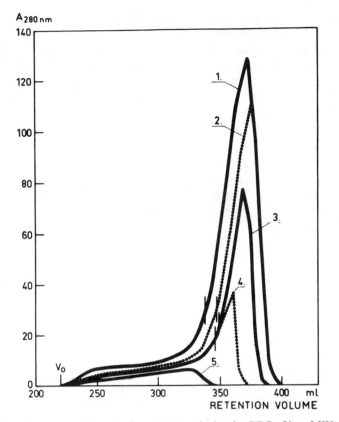

FIG. 8.15.   Influence of sample size on ion exclusion in GPC of low MW calcium lignosulphonate eluting from Sephadex G-25. Numbers 1 through 5 represent decreasing sample size. (Reproduced from Forss and Stelund[382] with permission of John Wiley and Sons.)

amounts injected only a very broad peak may be observed. Fortunately polyelectrolyte expansion, ion inclusion, and ion exclusion can all be eliminated by addition of electrolyte to the solvent. The effect of solvent electrolyte concentration and the volume and concentration of solutes have been examined in detail.[384] The use of GPC to determine Donnan equilibria has been described.[385]

Experimentally, both aqueous and organic solvent GPC are similar. The use of aqueous salt solutions raises concerns about corrosion of metal components. With pure water, the use of a conventional syphon requires modification[386] to produce uniform syphon dump volumes. This is achieved by using the effluent water to displace another more appropriate solvent which flows through the syphon. Calibration in aqueous systems is normally performed with Dextran fractions (Pharmacia Fine Chemicals) which are not narrow MWD standards. Sodium poly(styrene sulphonates) (Pressure Chemical Co.) have recently become available, and are produced by sulphonation of the narrow MWD polystyrene standards. The chemistry of this sulphonation has been described.[387] Calibration may also be performed by using a whole polymer of known molecular weight distribution.[388] The universal calibration approach was found to be valid for dextran and sodium poly(styrene sulphonates) in 0·2 M and 0·8 M sodium sulphonate with porous glass packings,[389] Fig. 8.16. The effect of solvent ionic strength with porous glass packings has been examined in detail.[381,390]

Recently, the major advance in aqueous GPC has been the development of new rigid column packings. This has led to both an increase in the number of applications and the speed of analysis. An exhaustive survey of these new column developments has appeared.[391] These improvements are now discussed on the basis of column packing type.

8.3.1. Porous Glasses and Silicas

These materials have found widespread use as aqueous GPC column packings in water and aqueous systems. The physical properties of porous glasses[65] and porous silica have been reviewed[66,392] and their use in GPC has been described.[393] These authors reported the use of a tetramethyl ammonium salt to prevent adsorption of non-ionic and cationic polymers in silica. Others have bonded a quaternary ammonium group onto the silica surface[394] and used acidic salt solutions to elute dextrans and polycations. Treatment of silica by proprietary method (Vit-X) has yielded materials suitable for aqueous application.[395]

Corning Porous Glass (CPG) has been used extensively in aqueous applications.[389,390,396] Its use has been extended by the use of surface treatment with poly(ethylene oxide) (glycol) of approximately 20 000 MW. Various biological[397] and virus samples have been successfully chromatographed by this modification. It has been shown that although this treatment suppresses the adsorption of proteins, ionic exclusion effects remain.[399] Protein adsorption has also been prevented by use of an amino acid buffer.[400] Proteins have also been chromatographed in CPG columns by use of their sodium dodecyl sulphate (SDS) complexes.[401,402] The latter authors reported the addition of 6 M urea to prevent adsorption.

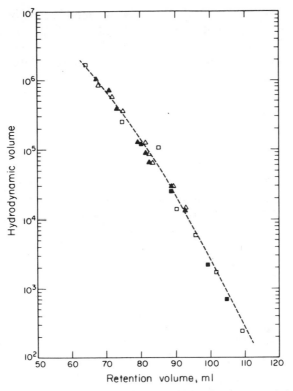

FIG. 8.16.   Validity of universal calibration for poly(styrene sulphonates) and dextran at different solvent ionic strengths.[389] ■, sodium polystyrene sulphonate—0·8 M sodium sulphate; ▲, dextran—0·8 M sodium sulphate; □, sodium polystyrene sulphonate—0·2 M sodium sulphate; △, dextran—0·2 M sodium sulphate. (Reproduced from Spatorico and Beyer[389] with permission of John Wiley and Sons.)

8.3.2. Derivatised Porous Glasses or Silicas

8.3.2.1. Glycophase CPG Column Packings

This glycerol type material is chemically bonded to the glass to improve the stability over the 'coating' approaches. They have been found useful for the characterisation of proteins in phosphate buffers[403] or denaturing solvents, viz. 6 M guanidine hydrochloride,[404] industrial glues,[403] and a polycation, chitosan.[405] The use of small particle size glass led to very efficient separations of proteins.[406,407] However, for poly(vinyl alcohol) untreated glass was found to be preferable.[403] Other successful applications of glycophase CPG have included[408] the water soluble celluloses (carboxy-methyl, hydroxyethyl, and carboxyethyl hydroxyethyl). For this application high ionic strength buffer 0·7 m, pH 3·7 and extremely small sample sizes, 50 μl were required. A method to determine soluble polymer in aqueous acrylamide at levels down to 5 ppm has been reported.[409]

Other bonded stationary phases have been evaluated[410] in an extensive study of materials suitable for aqueous GPC. Recently, a diol packing has been applied to the high performance GPC of proteins.[411,412]

8.3.3. TSK Column Packings

These materials have been widely used and may be considered as the most universal aqueous GPC column packing. Two types of TSK gel are available: type PW[413] and type SW.[414,415] The former is a cross-linked polymer and the latter is based on silica with a hydrophilic bonded coating. They are available in a variety of pore sizes, and their performance has been compared.[416] The PW type is more suitable for determination of MWD having a higher separating range and is more suitable for the separation of oligomers.[417] The SW type is more efficient for separation of proteins and has a high resolution in its separating range. The high-speed chromatography of proteins has been reported for TSK gel, type PW with phosphate buffer[418] and type SW with 6 m guanidine hydrochloride,[419] SDS[420,421] solvents and phosphate buffers.[422] The PW type material has been used to characterise poly(vinyl alcohol) fractions[423] (Fig. 8.17).

The PW type material has also been used to separate polycations, polyethyleneimine[424] using 0·1–0·4 m ethanolamine/acetic acid in aqueous medium at pH 5·1.

8.3.4. Bondagel and I-125 Column Packings

Waters Associates have introduced two modified silica based materials. One was a bonded polyether which can be used in aqueous and non-aqueous systems. Its use in aqueous systems for characterising poly(styrene sulphonate), poly(vinyl alcohol) and human plasma systems has been described.[425] The effect of solvents, low or intermediate ionic strength and high chaotropic property (concentrated urea) have been investigated[426] for these packing materials with proteins and nucleic acids. The more hydrophobic solutes required the presence of urea to prevent adsorption.

The other introduction was the I-125 protein column. The comparison of TSK-gel type SW, μ Bondagel and the I-125 protein columns for GPC of insulin has been published.[427] The μ Bondagel column required the

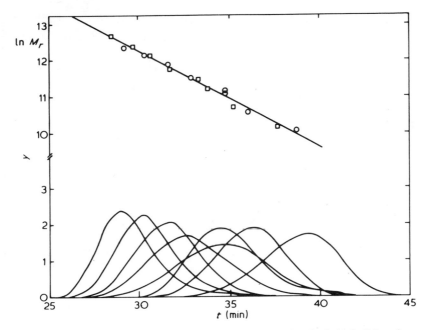

FIG. 8.17.   Chromatograms and calibration curve for poly(vinyl alcohol) fractions using TSK gel type PW. (Reproduced from Atkinson *et al.*[423] with permission of ICP Business Press.)

addition of detergent, urea, and acid to elute insulin properly. The TSK gel and the I-125 columns eluted insulin normally in either acidic or neutral conditions, with urea.

8.3.5. Enzacryl gels
Cross-linked poly(acryloylmorpholines) have been prepared as matrices for GPC[428] and used for aqueous applications.[429,430] These materials are capable of high resolution in the lower molecular weight range; Fig. 8.18 shows the chromatograms of poly(ethylene glycol) in water.

8.3.6. Other Hydrophilic Column Packings
Poly(hydroxy ethyl methacrylate) gels are available[431] under the trade name Spheron.[432] The high aliphatic content introduced by the cross-linking agent, ethylene dimethacrylate, leads to significant hydrophobic interactions. Shodex OH pak[433] is presumed to have a methacrylate glycerol structure,[434] again hydrophobic interaction is observed.

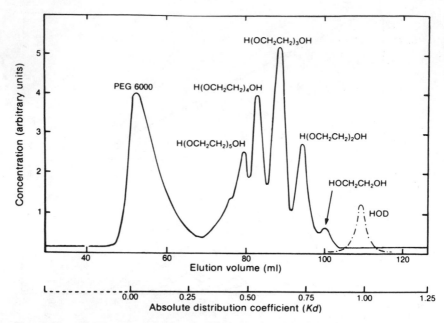

FIG. 8.18. Chromatogram of poly(ethylene glycol) using enzacryl gel K0 in water. (Reproduced from Epton[430] with permission of Aldrich Chemical Co.)

Sulphonated poly(styrene divinylbenzene) packings are available, e.g. Shodex Ionpack.[433] Application to carbohydrate and poly(ethylene glycol) analysis has been described.[435] Poly(vinyl alcohol) gels have been prepared,[436] dextrans, poly(ethylene oxides) and proteins have been successfully chromatographed. Styragel itself has been used for aqueous GPC of dextrans by the addition of 0.1% sodium lauryl sulphate to the solvent.[437]

8.3.7. Soft Gels for Aqueous GPC
Many gels are available in this category, all are capable of high resolution but can only withstand modest pressures. Analysis times are therefore long.

8.3.7.1. Agarose Gels
These materials are available under the trade names Sepharose (Pharmacia Fine Chemicals) or Bio-Gel A (Bio-Rad Laboratories). These gels may only be used to 36 °C because of melting of the gel. The molecular weight exclusion limits are high and they have been used to characterise gelatins[438] and

proteins and other macromolecules.[439] For the former 0·1 M urea/0·1 M sodium chloride was required for reproducible results, the latter employed 0·15 M sodium chloride or 1 M sodium chloride together with 5 mM diemal sodium buffer and 0·02% sodium azide. In order to improve the temperature and mechanical stability, agarose has been cross-linked with 2,3-dibromopropanol, desulphated and reduced. The product Sepharose CL may be heated to 120 °C without change in the gel structure. Denaturing solvents such as 6 M guanidine hydrochloride may be used for the characterisation of proteins.[440] Ultrogel (LKB Instruments Inc.) is a composite gel in which agarose is held within pores formed by polyacrylamide. They have a separating range greater than polyacrylamide gels themselves but again are limited to temperatures below 36 °C. A comparison of Sepharose, Sepharose CL, and Ultrogel for the characterisation of dextran has been published.[441]

8.3.7.2. Sephacryl Gels

These gels are produced by cross-linking allyl dextran with N,N-methylene-bisacrylamide. Its separating range lies between Sepharose and Sephadex. Its superior rigidity allows faster flow rates than either. Above pH 3 at moderate ionic strength proteins are eluted, but below pH 3 they bind strongly.[442] This material has also been used to fractionate water soluble proteins in formamide.[443]

8.3.7.3. Dextran Gels

Cross-linked dextran (Sephadex) gels are well known and widely used for a variety of solutes. In certain cases adsorption occurs with proteins as the pH is varied. For example, poly(L-histidine) elutes normally in pH 2·35 buffer but is adsorbed at pH 5·4.[446] It has been widely used for the separation of plasma proteins including preparative scale operation.[445] The dextran packings are compatible with pyridine and this has been exploited to extend their application to materials not soluble in aqueous media.[446,447] The application of Sephadex and Sepharose for the chromatographic characterisation of polysaccharides has been described and some non-ideal effects have been found for certain samples.[448]

8.3.7.4. Acrylamide Gels

Gels based on polyacrylamide are commercially available under the trade name Bio-Gel P (Bio-Rad Laboratories). These are soft gels and are capable of high resolution at the expense of high speed. They have been extensively used for oligosaccharide characterisation and the operational variables for this have been examined.[499]

8.3.7.5. Cellulose Gels

Hydroxyethyl cellulose gels have been described[450] for the characterisation of oligosaccharides and poly(ethylene oxides). These are of higher mechanical rigidity than the commercial soft gels, and in comparison with dextran and acrylamide-based materials are less adsorptive.

8.4. APPLICATIONS OF GEL PERMEATION CHROMATOGRAPHY FOR LOW MOLECULAR WEIGHT COMPOUNDS

Prior to the introduction of high efficiency columns based on micro particle ($< 10\ \mu$m) packing materials, the application of GPC for low molecular weight compounds was very limited. Compared with high pressure liquid chromatography (HPLC) however, it has the attractiveness of simplicity. With the advent of a variety of commercially available high efficiency columns, the application of GPC to low molecular weight additives and oligomers has increased significantly.

One of the attractive features is that with appropriate calibration, molecular weights would be determined rapidly in the oligomer region. Early workers correlated elution volumes with molar volume[451–452] rather than molecular weight or linear dimensions. This approach has been explored further[453–455] while others have proposed an effective chain length parameter.[456,457]

Solvent polarity is a critical parameter in the elution of oligomers. Polar molecules associate with tetrahydrofuran by hydrogen bonding.[453,458] Equations relating the molecular weight of oligomeric polystyrenes, poly-(ethylene glycol), epoxy resins, and PCR Novolaks the molecular weight of n-hydrocarbons through their elution volumes in THF and chloroform have been reported.[459] More sophisticated approaches attempt to correlate solute geometry with elution volume.[460–462]

The choice of solvent for GPC at low molecular weights is important.[463] For hydrocarbons with Styragel column packing, those solvents whose solubility parameter was closest to that of the gel gave the lowest distribution coefficients. Polar solvents such as acetone exhibit larger values of the distribution coefficient and the steric exclusion effect was no longer predominant. Similar investigations have been reported for low MW polystyrene solutes with various solvents and porous silica column packings.[464]

Reviews of applications of GPC to low molecular weight compounds in commercial polymers have appeared.[465-469] TSK gels[465] have been used to analyse sulphur from rubber samples and additives from polyolefins. Various oligomers have been separated using Du Pont SE columns.[466] μ Styragel columns have been used for the preparative separation of polymer extracts for subsequent analysis.[467] Shodex columns have been used to separate epoxy resins, n-paraffins, fatty acids, triglycerides, oligomers, oils, polymer additives, and adhesives.[468] Separations of polystyrene and poly-(ethylene glycol) oligomers and phthalate ester isomers have been reported using 5 μm TSK gel.[469]

The determination of molecular weight of poly(ethylene glycols) has been investigated by a variety of techniques.[470] Chloroform was used for the solvent and the three μ Styragel columns employed covered the MW range 150 to 20 000. Diallyl phthalate prepolymers[471] and low molecular weight compounds in poly(acryl ethers)[472] have been characterised by GPC with combined polystyrene gel–THF systems.

Often the chromatograms of oligomers contain regions at low molecular weight of discrete peaks and regions of higher molecular weight of fused peaks. Various methods for the calculation of molecular weight averages in each system have been examined.[221] The preferred method was to calculate the areas of the discrete peaks and divide the area of fused peaks at 0·25-min intervals and measure the peak heights.

REFERENCES

1. P. Flodin and B. Ingleman, US Patent 3,042,667, 1959.
2. J. C. Moore, *J. Polym. Sci.*, **A2**, 835, 1964.
3. H. Determann, *Gel Chromatography*, Springer-Verlag, NY, 1968.
4. K. H. Altgelt and L. Segal (Eds.), *Gel Permeation Chromatography*, Marcel Dekker, Inc., NY, 1971.
5. R. Epton (Ed.), *Chromatography of Synthetic and Biological Polymers*, Vol. I, *Column Packings, GPC, GF and Gradient Elution*, Ellis Horwood Ltd, Chichester, UK, 1978.
6. T. Kremmer and L. Boross, *Gel Chromatography. Theory, Methodology, Applications*, Wiley Interscience, NY, 1978.
7. J. Cazes (Ed.), *Liquid Chromatography of Polymers and Related Materials I*, Marcel Dekker, Inc., NY, 1977.
8. J. Cazes and X. Delamare (Eds.), *Liquid Chromatography of Polymers and Related Materials II*, Marcel Dekker, Inc., NY, 1980.

9. J. Cazes (Ed.), *Liquid Chromatography of Polymers and Related Materials III*, Marcel Dekker, Inc., NY, 1981.
10. W. W. Yau, J. J. Kirkland and D. D. Bly, *Modern Size-Exclusion Liquid Chromatography*, Wiley Interscience, NY, 1979.
11. T. Provder (Ed.), *Size Exclusion Chromatography (GPC)*, ACS Symposium Series No. 138, American Chemical Society, Washington, DC, 1980.
12. T. S. Work and E. Work (Eds.), *Laboratory Techniques in Biochemistry and Molecular Biology, An Introduction to Gel Chromatography*, L. Fischer, North-Holland/American Elsevier, Amsterdam, 1974.
13. L. H. Tung and J. C. Moore, Gel Permeation Chromatography, in *Fractionation of Synthetic Polymers*, Marcel Dekker, Inc., NY, 1977.
14. B. W. Hatt, Polymer Molecular Weight Distribution by Gel Permeation Chromatography, in *Developments in Chromatography—1*, C. E. H. Knapman (Ed.), Applied Science Publishers Ltd, London, 1978.
15. K. H. Altgelt, Gel Permeation Chromatography (GPC), in *Chromatography in Petroleum Analysis*, K. H. Altgelt and T. H. Gouw (Eds.), Marcel Dekker, Inc., NY, 1979.
16. S. G. Perry, R. Amos and P. I. Brewer, *Practical Liquid Chromatography*, A Plenum/Rosetta Edition, NY, 1972, Ch. 5.
17. N. C. Billingham, *Molar Mass Measurements in Polymer Science*, John Wiley and Sons, NY, 1977, Ch. 8.
18. J. V. Dawkins and G. Yeadon, High Performance Gel Permeation Chromatography, in *Developments in Polymer Characterisation—1*, J. V. Dawkins (Ed.), Applied Science Publishers Ltd, London, 1978, Ch. 3.
19. D. M. W. Anderson, Gel Permeation Chromatography, in *Practical High Performance Liquid Chromatography*, C. F. Simpson (Ed.), Heyden and Son Ltd, London, 1976, Ch. 9.
20. N. C. Billingham, Characterization of High Polymers by Gel Permeation Chromatography, in *Practical High Performance Liquid Chromatography*, C. F. Simpson (Ed.), Heyden and Son Ltd, London, 1976, Ch. 10.
21. P. C. Allen, E. A. Hill and A. M. Stokes, *Plasma Proteins*, Blackwell Scientific Publications, London, 1978.
22. K. K. Unger, Silica as a Packing in Size-Exclusion Chromatography, Porous Silica, in *Journal of Chromatography Library Volume 16*, Elsevier Scientific Publishing Co., Amsterdam, 1979.
23. K. H. Altgelt and J. C. Moore, Gel Permeation Chromatography, in *Polymer Fractionation*, M. J. R. Cantow (Ed.), Academic Press, NY, 1967, Ch. B4.
24. M. J. R. Cantow and J. F. Johnson, Gel Permeation Chromatography, in *Guide to Modern Methods of Instrumental Analysis*, T. H. Gouw (Ed.), John Wiley and Sons, Inc., NY, 1972, Ch. IV.
25. T. C. Laurent, B. Obrink, K. Hellsing and A. Wasteson, On the Theoretical Aspects of Gel Chromatography, in *Modern Separation Methods of Macromolecules and Particles*, T. Gerritsen (Ed.), Interscience, NY, 1969.
26. J. F. Johnson and R. S. Porter, Gel Permeation Chromatography, in *Progress in Polymer Science*, Vol. 2, A. D. Jenkins (Ed.), Pergamon Press, Oxford, 1970.
27. A. C. Ouano, E. M. Barrall II and J. F. Johnson, Gel Permeation Chromatography, in *Techniques and Methods of Polymer Evaluation*, 4(2), 287, 1975, Ch. 6.

28. A. C. Ouano, Kinematics of Gel Permeation Chromatography, in *Advances in Chromatography*, Vol. 15, J. C. Giddings (Ed.), Marcel Dekker, Inc., NY, 1977.

29. A. C. Ouano, Quantitative Data Interpretation Techniques in Gel Permeation Chromatography, *Rev. Macromol. Chem.*, **9**, 123, 1973.

30. K. H. Altgelt, Theory and Mechanics of Gel Permeation Chromatography, in *Advances in Chromatography*, Vol. 7, J. C. Giddings and R. A. Keller (Eds.), Marcel Dekker, Inc., NY, 1968, p. 3.

31. A. R. Cooper, J. F. Johnson and R. S. Porter, Gel Permeation Chromatography: Current Status, *Amer. Lab.*, p. 12, May 1973.

32. A. R. Cooper, Gel Permeation Chromatography, *Chem. Brit.*, **9**, 112, 1973.

33. P. L. Dubin, Polymer Separation, *Ind. Res.*, **55**, July 1976.

34. M. R. Ambler, Recent Advances in Polymer Characterization by GPC, *Amer. Lab.*, p. 16, May 1979.

35. R. A. Ellis, A Review of Gel Permeation Chromatography Parts I–III (respectively), *Pigm. Resin. Technol.*, **8**(9), 10, 1979; **8**(10), 4, 1979; **8**(11), 17, 1979.

36. J. Janca, Polymer Analysis by Size Exclusion Chromatography, *J. Liquid Chrom.*, **4** (Suppl. 1), 1, 1981.

37. R. E. Majors, High Performance Liquid Chromatography Packings and Columns, *Amer. Lab.*, 13, October 1975.

38. R. E. Major, Recent Advances in High Performance Liquid Chromatography Packings and Columns, *J. Chrom. Sci.*, **15**, 334, 1977.

39. V. F. Gaylor and H. L. James, Gel Permeation Chromatography (Liquid Exclusion Chromatography), *Anal. Chem. Rev.*, **48**(5), 44R, 1976.

40. V. F. Gaylor and H. L. James, Gel Permeation Chromatography (Steric Exclusion Chromatography), *Anal. Chem. Rev.*, **50**, 29R, 1978.

41. E. F. Casassa, *J. Phys. Chem.*, **75**, 3929, 1971.

42. E. F. Casassa, *J. Polym. Sci.*, **B5**, 773, 1967.

43. E. F. Casassa, *Separ. Sci.*, **6**, 305, 1971.

44. E. F. Casassa and Y. Tagami, *Macromol.*, **2**, 4, 1969.

45. E. F. Casassa, *Macromol.*, **9**, 182, 1976.

46. J. C. Giddings, E. Kucera, C. P. Russell and M. N. Myers, *J. Phys. Chem.*, **72**, 4397, 1968.

47. G. K. Ackers, *Biochem.*, **3**, 723, 1964.

48. W. W. Yau and D. P. Malone, *J. Polym. Sci.*, **B5**, 663, 1967.

49. E. Z. Di Marzio and C. M. Guttman, *J. Polym. Sci.*, **7**, 267, 1969.

50. W. W. Yau, C. P. Malone and S. W. Fleming, *J. Polym. Sci.*, **B6**, 803, 1968.

51. W. W. Yau, *J. Polym. Sci.*, *A2*, **7**, 483, 1969.

52. M. E. Van Kreveld and N. Van Der Hoed, *J. Chromatog.*, **83**, 111, 1973.

53. W. W. Yau, J. J. Kirkland, D. D. Bly and H. J. Stoklosa, *J. Chromatogr.*, **125**, 219, 1976.

54. M. E. Van Kreveld and N. Van Den Hoed, *J. Chromatogr.*, **149**, 71, 1978.

55. J. H. Aubert and M. Tirrell, *Separ. Sci. and Technol.*, **15**, 123, 1980.

56. J. H. Aubert and M. Tirrell, *Rheol Acta.*, **19**, 452, 1980.

57. J. V. Dawkins, *J. Liquid Chromatogr.*, **1**, 279, 1978.

58. J. V. Dawkins, *Polymer*, **19**, 705, 1978.

59. J. V. Dawkins, *Pure and Appl. Chem.*, **51**, 1473, 1979.

60. R. Audebert, *Polymer*, **20**, 1561, 1979.
61. J. Lecourtier, R. Audebert and C. Quivoron, *Pure and Appl. Chem.*, **51**, 1483, 1979.
62. B. G. Belenkii, *Pure and Appl. Chem.*, **51**, 1519, 1979.
63. ASTM D3536-76, ASTM 1916 Race St., Philadelphia, PA 19103.
64. A. R. Cooper and A. R. Bruzzone, *J. Polym. Sci.*, *A2*, **11**, 1423, 1973.
65. A. R. Cooper, A. R. Bruzzone, J. H. Cain and E. M. Barrall II, *J. Appl. Polym. Sci.*, **15**, 571, 1971.
66. A. R. Cooper and E. M. Barrall II, *J. Appl. Polym. Sci.*, **17**, 1253, 1973.
67. A. R. Cooper, in *Chromatography of Synthetic and Biological Polymers*, Vol. 1, R. Epton (Ed.), Ellis Horwood Ltd, Chichester, UK, 1978, p. 344.
68. W. W. Yau, C. R. Ginnard and J. J. Kirkland, *J. Chromatogr.*, **149**, 465, 1978.
69. D. Ishii, K. Hibi, K. Asai, T. Jonokuchi and M. Nagaya, *J. Chromatog.*, **144**, 157, 1977.
70. D. Ishii, K. Hibi, K. Asai and T. Jonokuchi, *J. Chromatogr.*, **151**, 147, 1978.
71. E. M. Barrall II, M. J. R. Cantow and J. F. Johnson, *J. Appl. Polym. Sci.*, **12**, 1373, 1968.
72. S. Mori, *Anal. Chem.*, **50**, 1639, 1978.
73. K. Saitoh and N. Suzuki, *Anal. Chem.*, **51**, 1683, 1979.
74. G. E. James, Hewlett-Packard Co. UV/VIS Technical Papers UV-2, 1980.
75. J. H. Ross and M. E. Castro, *J. Polym. Sci.*, **C21**, 143, 1968.
76. S. L. Terry and F. Rodriguez, *J. Polym. Sci.*, **C21**, 991, 1968.
77. A. R Cooper and J. F. Johnson, *J. Appl. Polym. Sci.*, **13**, 1487, 1969.
78. F. M. Mirabella, Jr., E. M. Barrall II and J. F. Johnson, *Polymer*, **17**, 17, 1976.
79. E. G. Bartick, *J. Chromatog. Sci.*, **17**, 336, 1979.
80. D. W. Vidrine and D. R. Matteson, *Appl. Spectrosc.*, **32**, 502, 1978.
81. D. W. Vidrine, *J. Chromatogr. Sci.*, **17**, 477, 1979.
82. J. Francois, M. Jacot, Z. Grubisic-Gallot and H. Benoit, *J. Appl. Polym. Sci.*, **22**, 1159, 1978.
83. D. Sarazin, J. LeMoigne and J. Francois, *J. Appl. Polym. Sci.*, **22**, 1377, 1978.
84. H. Leopold and B. Trathnigg, *Die Angew. Makromol. Chemie.*, **68**, 185, 1978.
85. B. Trathnigg, *Monatsh. Chem.*, **109**, 467, 1978.
86. B. Trathnigg, *Die Angew. Makromol. Chemie.*, **89**, 65, 1980.
87. B. Trathnigg, *Die Agnew. Makromol. Chemie.*, **89**, 73, 1980.
88. B. Trathnigg, *Makromol. Chem. Rapid Comm.*, **1**, 569, 1980.
89. W. W. Shulz and W. H. King, Jr., *J. Chromatogr. Sci.*, **11**, 343, 1973.
90. C. E. M. Morris and I. Grabovac, *J. Chromatogr.*, **189**, 259, 1980.
91. R. M. Screaton, P. F. Cullen, R. W. Seemann and M. D. Saunders, *J. Polym. Sci., Symp. No. 43*, 311, 1973.
92. D. O. Gray, *J. Chromatogr.*, **37**, 320, 1968.
93. D. D. Bly, W. W. Yau and H. J. Stoklosa, *Anal. Chem.*, **48**, 1256, 1976.
94. D. D. Bly, H. J. Stoklosa, J. J. Kirkland and W. W. Yau, *Anal. Chem.*, **47**, 1810, 1975.
95. G. A. Moebus, J. A. Crowther, E. G. Bartick and J. F. Johnson, *J. Appl. Polym. Sci.*, **23**, 3501, 1979.
96. H. E. Pickett, M. J. R. Cantow and J. F. Johnson, *J. Appl. Polym. Sci.*, **10**, 917, 1966.

97. J. A. Hassell, F. A. Sliemers, E. Drauglis and G. P. Nance, *J. Polym. Sci.*, *Polym. Lett. Edn.*, **17**, 111, 1979.
98. L. Füzes, *J. Appl. Polym. Sci.*, **24**, 405, 1979.
99. J. C. Giddings, L. M. Bowman and M. N. Myers, *Makromol.*, **11**, 443, 1977.
100. A. C. Ouano, *Advances in Chromatography*, Vol. 15, J. C. Giddings (Ed.), Marcel Dekker, NY, 1977, p. 233.
101. Y. Kato, S. Kido, M. Yamamoto and T. Hashimoto, *J. Polym. Sci.*, *Polym. Phys. Edn.*, **12**, 1339, 1974.
102. M. R. Ambler, L. J. Fetters and Y. Kesten, *J. Appl. Polym. Sci.*, **21**, 2439, 1977.
103. S. Mori, *J. Appl. Polym. Sci.*, **21**, 1921, 1977.
104. A. Campos, L. Borque and J. E. Figueruelo, *Anales de Quimica*, **74**, 701, 1978.
105. A. K. Mukherji, *J. HRCC and CC*, 273, November 1978.
106. J. V. Dawkins, T. Stone and G. Yeadon, *Polymer*, **18**, 1179, 1977.
107. J. V. Dawkins, T. Stone, G. Yeadon and F. P. Warmer, *Polymer*, **20**, 1165, 1979.
108. Y. Kato, S. Kido, H. Watanabe, M. Yamamoto and T. Hashimoto, *J. Appl. Polym. Sci.*, **19**, 629, 1975.
109. Y. Kato, S. Kido and T. Hashimoto, *J. Polym. Sci.*, *Polym. Phys. Edn.*, **11**, 2329, 1973.
110. J. Janca, *Advances in Chromatography*, Vol. 19, J. C. Giddings, E. Grushka, J. Cazes and P. R. Brown (Eds.), Marcel Dekker, Inc., NY, 1981, p. 37.
111. H. L. Berger and A. R. Schultz, *J. Polym. Sci.*, *Pt A*, **3**, 3643, 1965.
112. H. Coll and D. K. Gilding, *J. Polym. Sci.*, *Pt A-2*, **8**, 89, 1970.
113. C. V. Linden, *Polymer*, **21**, 171, 1980.
114. S. Mori and T. Suzuki, *J. Liquid Chromatogr.*, **3**, 343, 1980.
115. M. J. R. Cantow, R. S. Porter and J. F. Johnson, *J. Polym. Sci.*, *A-1*, **5**, 1391, 1967.
116. A. H. Abdel-Alim and A. E. Hamielec, *J. Appl. Polym. Sci.*, **18**, 297, 1974.
117. J. N. Cardenas and K. F. O'Driscoll, *J. Polym. Sci.*, *Polym. Lett. Edn.*, **13**, 657, 1975.
118. J. R. Purdon, Jr. and R. D. Mate, *J. Polym. Sci.*, *A-1*, **6**, 243, 1968.
119. B. R. Log, *J. Polym. Sci.*, *Polym. Chem. Ed.*, **14**, 2321, 1976.
120. T. D. Swartz, D. D. Bly and A. S. Edwards, *J. Appl. Polym. Sci.*, **16**, 3353, 1972.
121. P. Crouzet, A. Manteville, M. Lucet and A. Mortens, *Analysis*, **4**, 450, 1976.
122. I. B. Tsvetkovskii, V. I. Valuev and R. A. Shlyakhter, *Vysokomol. Soedin. Ser. A*, **19**, 2637, 1977.
123. W. W. Yau, H. J. Stoklosa and D. D. Bly, *J. Appl. Polym. Sci.*, **21**, 1911, 1977.
124. R. R. Vrijbergen, A. S. Soeteman and J. A. M. Smit, *J. Appl. Polym. Sci.*, **22**, 267, 1978.
125. F. L. McCrackin, *J. Appl. Polym. Sci.*, **21**, 191, 1977.
126. R. P. Chaplin, J. K. Haken and J. J. Paddon, *J. Chromatogr.*, **171**, 55, 1979.
127. M. J. Pollock, J. F. MacGregor and A. E. Hamielec, *J. Liquid Chromatogr.*, **2**, 895, 1979.
128. H. Benoit, Z. Grubisic, P. Rempp, D. Decker and J. G. Zilliox, *J. Chem. Phys.*, **63**, 507, 1966.

129. J. V. Dawkins, *Brit. Polym. J.*, **4**, 87, 1972.
130. J. V. Dawkins, *Europ. Polym. J.*, **13**, 837, 1977.
131. G. Samay, M. Kubin and J. Podesva, *Die Angew. Makromol. Chemie.*, **72**, 185, 1978.
132. J. V. Dawkins and M. Hemming, *Polymer*, **16**, 554, 1975.
133. A. R. Weiss and E. Cohn-Ginsberg, *J. Polym. Sci., Polym. Lett. Edn.*, **7**, 379, 1969.
134. G. Samay, M. Kubin and J. Podesva, *Die Angew. Makromol. Chemie.*, **72**, 185, 1978.
135. A. E. Hamielec and A. C. Ouano, *J. Liquid. Chromatogr.*, **1**, 111, 1978.
136. R. D. Hester and P. H. Mitchell, *J. Polym. Sci., Polym. Chem. Edn.*, **18**, 1727, 1980.
137. A. Rudin, *J. Polym. Sci., A1*, **9**, 2587, 1971.
138. A. Rudin and H. C. W. Hoegy, *J. Polym. Sci., A1*, **10**, 217, 1972.
139. D. Berek, D. Bakos, L. Soltes and T. Bleha, *J. Polym. Sci., Polym. Lett. Edn.*, **12**, 277, 1974.
140. A. Rudkin and R. A. Wagner. *J. Appl. Polym. Sci.*, **20**, 1483, 1976.
141. H. K. Mahabadi and A. Rudin, *Polymer J.*, **11**, 123, 1979.
142. J. Janca, *J. Chromatogr.*, **134**, 263, 1977.
143. J. Janca and S. Pokorny, *J. Chromatog.*, **148**, 31, 1978.
144. J. Janca and S. Pokorny, *J. Chromatog.*, **156**, 27, 1978.
145. J. Janca, *J. Chromatogr.*, **170**, 309, 1979.
146. J. Janca, *J. Chromatogr.*, **170**, 319, 1979.
147. J. Janca, *J. Chromatogr.*, **187**, 21, 1980.
148. J. Janca, *Anal. Chem.*, **51**, 637, 1979.
149. T. Bleha, D. Bakos and D. Berek, *Polymer*, **18**, 897, 1977.
150. T. Bleha, J. Mlynek and D. Berek, *Polymer*, **21**, 798, 1980.
151. J. E. Figueruelo, V. Soria and A. Campos, *Die Makromol. Chemie.*, **180**, 1069, 1979.
152. A. C. Ouano, *J. Polym. Sci., A1*, **10**, 2169, 1972.
153. D. Goedhart and A. Opschoor, *J. Polym. Sci., A2*, **8**, 1227, 1970.
154. Z. Grubisic-Gallot, M. Picot, P. Gramain and H. Benoit, *J. Appl. Polym. Sci.*, **16**, 2931, 1972.
155. A. L. Spatorico and B. Coulter, *J. Polym. Sci., Polym. Phys. Edn.*, **11**, 1139, 1973.
156. J. Janca and M. Kolinsky, *J. Chromatogr.*, **132**, 187, 1977.
157. J. Janca and S. Pokorny, *J. Chromatogr.*, **134**, 273, 1977.
158. W. S. Park and W. W. Graessley, *J. Polym. Sci., Polym. Phys. Edn.*, **15**, 71, 1977.
159. W. S. Park and W. W. Graessley, *J. Polym. Sci., Polym. Phys. Edn.*, **15**, 85, 1977.
160. T. Kotaka, *J. Appl. Polym. Sci.*, **21**, 501, 1977.
161. K. C. Berger, *Die Makromol. Chemie.*, **179**, 719, 1978.
162. A. C. Ouano and W. Kaye, *J. Polym. Sci., Polym. Chem. Edn.*, **12**, 1151, 1974.
163. A. C. Ouano, *J. Chromatogr.*, **118**, 303, 1976.
164. M. L. McConnell, *Amer. Lab.*, **63**, May 1978.
165. R. C. Jordan, *J. Liquid Chromatogr.*, **3**, 439, 1980.
166. Y. B. MacRury amd M. L. McConnell, *J. Appl. Polym. Sci.*, **24**, 651, 1979.

167. L. H. Tung, *J. Appl. Polymer. Sci.*, **10**, 375, 1966.
168. N. Friis and A. Hamielec, *Adv. Chromatogr.*, **13**. 41, 1975.
169. A. E. Hamielec, *J. Liquid Chromatogr.*, **3**, 381, 1980.
170. L. Marais, Z. Gallot and H. Benoit, *J. Appl. Polym. Sci.*, **21**, 1955, 1977.
171. R. V. Figini, *Polymer Bull.*, **1**, 619, 1979.
172. A. E. Hamielec and W. H. Ray, *J. Appl. Polym. Sci.*, **13**, 1319, 1969.
173. T. Provder and E. M. Rosen, *Separ. Sci.*, **5**, 437, 1970.
174. L. H. Tung and J. C. Moore, Gel Permeation Chromatography, in *Fractionation of Synthetic Polymers*, L. H. Tung (Ed.), Marcel Dekker, Inc., NY, 1977, Ch. 6.
175. L. H. Tung and J. R. Runyon, *J. Appl. Polym. Sci.*, **13**, 2397, 1969.
176. J. L. Waters, *J. Polym. Sci.*, *Pt A2*, **8**, 411, 1970.
177. Z. Grubisic-Gallot, L. Marais and H. Benoit, *J. Polym. Sci.*, *Polym. Phys. Edn.*, **14**, 959, 1976.
178. F. L. McCrackin and H. L. Wagner, *Macromol.*, **13**, 685, 1980.
179. K. C. Berger, *Die Makromol. Chemie.*, **180**, 2567, 1979.
180. J. R. Runyon, D. E. Barnes, J. F. Rudd and L. H. Tung, *J. Appl. Polym. Sci.*, **13**, 2359, 1969.
181. P. R. Griffiths, in *Analytical Applications of FT-IR to Molecular and Biological Systems*, J. R. Durig (Ed.), D. Reidel Pub. Co., Dordrecht (Holland), 1980, p. 149.
182. B. G. Willis, Hewlett-Packard Co. UV/VIS Appl. Note AN 295–5, 1980.
183. N. Ho-Duc and J. Prud'homme, *Macromol.*, **6**, 472, 1973.
184. A. Revillon, *J. Liquid Chromatogr.*, **3**, 1137, 1980.
185. Z. Grubisic-Gallot, L. Marais and H. Benoit, *J. Chromatogr.*, **83**, 363, 1973.
186. F. S. C. Chang, *J. Chromatogr.*, **55**, 67, 1971.
187. L. H. Tung, *J. Appl. Polym. Sci.*, **24**, 953, 1979.
188. C. Booth, J-L. Forget, I. Georgii, W. S. Li and C. Price, *Europ. Polym. J.*, **16**, 255, 1980.
189. S. T. Balke and R. D. Patel, *J. Polym. Sci.*, *Polym. Lett. Edn.*, **18**, 453, 1980.
190. S. T. Balke and R. D. Patel, in *Size Exclusion Chromatography*, T. Provder (Ed.), *ACS Symp. Ser. No. 138*, 1980.
191. E. E. Drott, in *Chromatographic Science Series*, Vol. 8, *Liquid Chromatography of Polymers and Related Materials I*, J. Cazes (Ed.), Marcel Dekker, Inc., NY, 1977, p. 161.
192. C. Price, J. L. Forget and C. Booth, *Polymer*, **18**, 527, 1977.
193. M. R. Ambler, *J. Appl. Polym. Sci.*, **21**, 1655, 1977.
194. L. H. Tung, *J. Polym. Sci.*, *A2*, **7**, 47, 1969.
195. Th. G. Scholte and N. C. J. Meijerink, *Brit. Polym. J.*, **9**, 133, 1977.
196. D. Constantin, *Europ. Polym. J.*, **13**, 907, 1977.
197. W. Scheinert, *Die Angew. Makromol. Chemie.*, **63**, 117, 1977.
198. W. S. Park and W. W. Graessley, *J. Polym. Sci.*, *Polym Phys. Edn.*, **15**, 85, 1977.
199. L. Marais, Z. Gallot and M. Benoit, *Analysis*, **4**, 449, 1976.
200. A. Servotte and R. de Bruille, *Die Makromol. Chemie.*, **176**, 203, 1975.
201. H. L. Wagner and F. L. McCrackin, *J. Appl. Polym. Sci.*, **21**, 2833, 1977.
202. T. Kato, A. Kanda, A. Takahashi, I. Noda, S. Maki and M. Nagasawa, *Polym. J.*, **11**, 575, 1979.

203. A. E. Hamielec, A. C. Ouano and L. L. Nebenzahl, *J. Liquid Chromatogr.*, **1**, 527, 1978.
204. D. E. Axelson and W. C. Knapp, *J. Appl. Polym. Sci.*, **25**, 119, 1980.
205. A. R. Cooper, A. J. Hughes and J. F. Johnson, *J. Appl. Polym. Sci.*, **19**, 435, 1975.
206. A. R. Cooper, A. J. Hughes and J. F. Johnson, *Chromatographia*, **8**, 136, 1975.
207. M. F. Vaughan and R. Dietz, in *Chromatography of Synthetic and Biological Polymers*, Vol. 1, R. Epton (Ed.), Ellis Horwood Ltd, Chichester, UK, 1978, p. 199.
208. Y. Kato, T. Kametani, K. Furukawa and T. Hashimoto, *J. Polym. Sci.*, *Polym. Phys. Edn.*, **13**, 1695, 1975.
209. C. M. L. Atkinson and R. W. Dietz, *Polymer*, **18**, 408, 1977.
210. S. Hattori, H. Endoh, N. Nakahara, T. Kamata and M. Hamashima, *Polymer J.*, **10**, 173, 1978.
211. S. Hattori, M. Hamashima, N. Nakahara and T. Kamato, *Chem. High Polym. Japan*, **34**, 503, 1977.
212. M. F. Vaughan and M. A. Francis, *J. Appl. Polym. Sci.*, **21**, 2409, 1977.
213. K. Dodgson, D. Sympson and J. A. Semlyen, *Polymer*, **19**, 1241, 1978.
214. J. Springer, J. Schmelzer and T. Zeplichal, *Chromatographia*, **13**, 164, 1980.
215. W. Heitz and H. Ullner, *Die Makromol. Chemie.*, **120**, 58, 1968.
216. K. J. Bombaugh and R. F. Levagnie, *Separ. Sci.*, **5**, 751, 1970.
217. P. E. Barker, F. J. Ellison and B. W. Hatt, in *Chromatography of Synthetic and Biological Polymers*, Vol. 1, R. Epton (Ed.), Ellis Horwood Ltd, Chichester, UK, 1978, p. 218.
218. P. E. Barker, B. W. Hatt and A. N. Williams, *Chromatographia*, **10**, 377, 1977.
219. W. W. Yau and S. W. Fleming, *J. Appl. Polym. Sci.*, **12**, 2111, 1968.
220. S. W. Hawley, *Chromatographia*, **11**, 499, 1978.
221. S. Mori, *J. Chromatogr.*, **156**, 111, 1978.
222. L. A. Prince and M. E. Stapelfeldt, *Sixth International GPC Seminar*, Miami Beach, Seminar Proceedings, October 1968, p. 78.
223. M. A. Dudley, *J. Appl. Polym. Sci.*, **16**, 493, 1978.
224. T. D. Swartz, D. D. Bly and A. S. Edwards, **16**, 3353, 1972.
225. E. K. Walsh, *J. Chromatogr.*, **55**, 193, 1971.
226. P. S. Ede, *J. Chromatogr. Sci.*, **9**, 275, 1971.
227. G. Pastuska and U. Just, *Die Angew. Makromol. Chemie.*, **81**, 11, 1979.
228. E. E. Drott, in *Liquid Chromatography of Polymer and Related Materials I*, J. Cazes, (Ed.), Marcel Dekker, Inc., NY, 1977, p. 41.
229. T. Provder, J. C. Woodbrey and J. H. Clark, in *Gel Permeation Chromatography*, K. H. Altgelt and L. Segel (Eds), Marcel Dekker, Inc., NY, 1971, p. 493.
230. A. R. Cooper, unpublished results.
231. E. Jacobi, H. Schuttenberg and R. C. Schulz, *Makromol. Chem., Rapid Commun.*, **1**, 397, 1980.
232. M. Arpin and C. Strazielle, *Polymer*, **18**, 591, 1977.
233. G. C. Berry and S. P. Yen, *Adv. Chem. Ser. No. 91*, 734, 1969.
234. F. W. Billmeyer and I. Katz, *Makromolecules*, **2**, 105, 1969.
235. W. Y. Lee, *J. Appl. Polym. Sci.*, **22**, 3343, 1978.

236. M. Minarik, Z. Sir and J. Coupek, *Die Angew. Makromol. Chemie.*, **64**, 147, 1977.
237. J. R. Overton, J. Rash and L. D. Moore, Jr., *Sixth International GPC Seminar*, Miami Beach, Seminar Proceedings, October 1968, p. 422.
238. G. Shaw, *Seventh International GPC Seminar*, Monte Carlo, Seminar Proceedings, 1969, p. 309.
239. E. E. Paschke, B. A. Bidlingmeyer and J. G. Bergman, *J. Polym. Sci., Polym. Chem. Edn.*, **15**, 983, 1977.
240. Waters Associates, unpublished results.
241. E. L. Slagowski, R. C. Gebauer and G. J. Gaesser, *J. Appl. Polym. Sci.*, **21**, 2293, 1977.
242. S. Shiono, *J. Polym. Sci., Polym. Chem. Edn.*, **17**, 4123, 1979.
243. S. Shiono, *Anal. Chem.*, **51**, 2398, 1979.
244. A. W. Birley, J. V. Dawkins and D. Kyriacos, *Polymer*, **21**, 632, 1980.
245. H. Batzer and S. A. Zahir, *J. Appl. Polym. Sci.*, **19**, 585, 1975.
246. D. Braun and D. W. Lee, *Die Angew. Makromol. Chemie.*, **57**, 111, 1977.
247. J. M. Charlesworth, *J. Polym. Sci., Polym. Phys. Edn.*, **17**, 1571, 1979.
248. W. A. Dark, E. C. Conrad and L. W. Crossman, Jr., *J. Chromatogr. Sci.*, **91**, 247, 1974.
249. A. A. Wickham, D. D. Rice and R. J. Dubois, *24th SAMPE Symposium*, May 1979, p. 506.
250. G. L. Hagnauer, *Technical Report AMMRC, TR 79-59*, Army Materials and Mechanics Research Center, Waterdown, Ma 02172, November 1979.
251. W. Heitz, *Sixth International GPC Seminar*, Miami Beach, Seminar Proceedings, October 1968, p. 130.
252. J. W. Aldersley, V. M. R. Bertram, G. R. Harper and B. P. Stark, *Br. Polym. J.*, **1**, 101, 1969.
253. J. W. Aldersley and P. Hope, *Die Angew. Makromol. Chemie.*, **24**, 137, 1972.
254. E. R. Wagner and R. J. Greff, *J. Polym. Sci., A1*, **9**, 2193, 1971.
255. M. Duval, B. Bloch and S. Kohn, *J. Appl. Polym. Sci.*, **16**, 1585, 1972.
256. M. Cornia, G. Sartori, G. Casnati and G. Casiraghi, *J. Liquid Chromatogr.*, **4**, 13, 1981.
257. D. Braun and J. Arndt, *Die Angew. Makromol. Chemie*, **73**, 143, 1978.
258. P. Hope, R. Anderson and A. S. Bloss, *Br. Polym. J.*, **5**, 67, 1973.
259. R. W. Hemingway and G. W. McGraw, *J. Liquid Chromatogr.*, **1**, 163, 1978.
260. B. Feurer and A. Gourdenne, *Polym. Prepr.*, **15**(2), 279, 1974.
261. D. Braun and J. Arndt, *Fresenius Z. Anal. Chem.*, **294**, 130, 1979.
262. P. Hope, B. P. Stark and S. A. Zahir, *Br. Polym. J.*, **5**, 363, 1973.
263. J. E. Armonas, *Forest Prod. J.*, **20**, 22, 1970.
264. W. Dankelman, J. M. H. Daemen, A. J. J. deBreet, J. L. Mulder, W. G. B. Huysmans and J. deWit, *Die Angew. Makromol. Chemie*, **54**, 187, 1976.
265. D. Braun and F. Bayersdorf, *Die Angew. Makromol. Chemie*, **85**, 1, 1980.
266. K. Kumlin and R. Simonson, *Die Angew. Makromol. Chemie*, **68**, 175, 1978.
267. N. D. Hann, *J. Polym. Sci., Polym. Chem. Edn.*, **15**, 1331, 1977.
268. F. Spagnolo and W. M. Malone, *J. Chromatogr. Sci.*, **14**, 52, 1976.
269. P. McFadyen, *J. Chromatogr.*, **123**, 468, 1976.
270. C. L. Swanson, R. A. Buchanan and F. M. Otey, *J. Appl. Polym. Sci.*, **23**, 743, 1979.

271. W. V. Smith and S. Thirwengada, *Rubber Chem. and Technol.*, **43**, 1439, 1970.
272. M. R. Ambler, *J. Appl. Polym. Sci.*, **25**, 901, 1980.
273. W. V. Smith, *J. Appl. Polym. Sci.*, **18**, 3685, 1974.
274. J. N. Anderson, *J. Appl. Polym. Sci.*, **18**, 2819, 1974.
275. J. N. Anderson, S. K. Baczek, H. E. Adams and L. E. Vescelius, *J. Appl. Polym. Sci.*, **19**, 2255, 1975.
276. S. K. Baczek, J. N. Anderson and H. E. Adams, *J. Appl. Polym. Sci.*, **19**, 2269, 1975.
277. M. M. Coleman and R. F. Fuller, *J. Macromol. Sci.*, *Phys.*, **B11**, 419, 1975.
278. J. L. White, D. G. Salladay, D. O. Quisenberry and D. I. Maclean, *J. Appl. Polym. Sci.*, **16**, 2811, 1972.
279. E. N. Fuller, G. T. Porter and L. B. Roof, *J. Chromatog. Sci.*, **17**, 661, 1979.
280. H. E. Adams, in *Gel Permeation Chromatography*, K. H. Altgelt and L. Segal (Eds), Marcel Dekker, Inc., NY, 1971, p. 391.
281. K. F. Elgert and R. Wohlschiess, *Die Angew. Makromol. Chemie*, **57**, 87, 1977.
282. V. F. Gaylor, H. L. James and J. P. Herdering, *J. Polym. Sci.*, *Polym. Chem. Edn.*, **13**, 1575, 1975.
283. G. C. N. Cheesman and K. R. N. Pollard, *Internat. Polym. Latex Conf.*, London, November 1978.
284. J. V. Dawkins and G. Yeadon, *Polymer*, **20**, 981, 1979.
285. J. V. Dawkins and G. Yeadon, *J. Chromatogr. Sci.*, **188**, 333, 1980.
286. Y. Kato, S. Kido and T. Hashimoto, *J. Polym. Sci.*, *Polym. Phys. Edn.*, **11**, 2329, 1973.
287. L. J. Fetters, *J. Appl. Polym. Sci.*, **20**, 3437, 1976.
288. L. J. Fetters and D. McIntyre, *Polymer*, **19**, 463, 1978.
289. J. V. Dawkins and M. Hemming, *Die Makromol. Chemie*, **176**, 1777, 1975.
290. J. V. Dawkins and M. Hemming, *Die Makromol. Chemie*, **176**, 1795, 1975.
291. J. V. Dawkins and M. Hemming, *Die Makromol. Chemie*, **176**, 1815, 1975.
292. T. Spychaj and D. Berek, *Polymer*, **20**, 1108, 1979.
293. E. L. Slagowski, L. J. Fetters and D. McIntyre, *Macromolecules*, **7**, 394, 1974.
294. Y. Kato, T. Kametani and T. Hashimoto, *J. Polym. Sci.*, *Polym. Phys. Edn.*, **14**, 2105, 1976.
295. L. Soltes, D. Berek and D. Mikulasova, *Colloid and Polym. Sci.*, **258**, 702, 1980.
296. W. B. Smith and H. W. Temple, *J. Phys. Chem.*, **72**, 4613, 1968.
297. B. Wandelt, J. Brzezinski and M. Kryszewski, *Europ. Polym. J.*, **16**, 583, 1980.
298. T. Ogawa, M. Sakai and W. Ishitobi, *J. Liquid Chromatogr.*, **1**, 151, 1978.
299. S. Mori, *J. Chromatogr.*, **194**, 163, 1980.
300. W. Siebourg, R. D. Lundberg and R. W. Lenz, *Macromolecules*, **13**, 1013, 1980.
301. T. Provder, J. C. Woodbrey and J. H. Clark, *Separ. Sci.*, **6**, 101, 1971.
302. S. Hattori, M. Hamashima, H. Nakahara and T. Kamata, *Kobunshi Ronbunshu*, **34**, 503, 1977.
303. M. J. Bowden, *J. Polym. Sci.*, *Symp. No. 49*, 221, 1975.
304. E. Gipstein, A. C. Ouano, D. E. Johnson and O. U. Need III, *IBM J. Res. Dev.*, **21**, 143, 1977.

305. P. Vlcek, J. Janca and J. Trekoval, *Makromol. Chem., Rapid Commun.*, **1**, 485, 1980.
306. D. Gandhi, E. Ellis, A. Grugnola and E. Radin, University of Lowell, Ma., personal communication.
306a. N. Nakajima, in *Polymer Molecular Weight Methods*, M. Ezoin (Ed.), *Advances in Chemistry Series No. 125*, American Chemical Society, 1973, p. 98.
307. E. E. Drott and R. A. Mendelson, *J. Polym. Sci., Pt A2*, **8**, 1361, 1970.
308. E. E. Drott and R. A. Mendelson, *J. Polym. Sci., Pt A2*, **8**, 1373, 1970.
309. D. M. Sadler, T. Williams, A. Keller and I. M. Ward, *J. Appl. Polym. Sci., A2*, **7**, 1819, 1969.
310. R. C. Ferguson, H. J. Stoklosa, W. W. Yau and H. M. Hoehn, *J. Appl. Polym. Sci., Appl. Polym. Symp. No. 34*, 119, 1979.
311. P. Crouzet, F. Fine and P. Mangin, *J. Appl. Polym. Sci.*, **13**, 205, 1969.
312. C. M. L. Atkinson and R. Dietz, *Die Makromol. Chemie*, **177**, 213, 1976.
313. T. Ogawa and T. Inabe, *J. Appl. Polym. Sci.*, **21**, 2979, 1977.
314. J. Echarri, J. J. Irwin, G. M. Guzman and J. Ansorena, *Die Makromol. Chemie*, **180**, 2749, 1979.
315. R. Salovey and R. C. Gebauer, *J. Appl. Polym. Sci.*, **17**, 2811, 1973.
316. K. B. Andersson, A. Holmstrom and E. M. Sorvik, *Die Makromol. Chemie*, **166**, 247, 1973.
317. A. H. Abdel-Alim and A. E. Hamielec, *J. Appl. Polym. Sci.*, **17**, 3033, 1973.
318. C. M. L. Atkinson, R. Dietz and J. H. S. Green, *J. Macromol. Sci., Phys.*, **B14**, 101, 1977.
319. A. H. Abdel-Alim, *J. Appl. Polym. Sci.*, **23**, 1577, 1979.
320. A. Peyrouset and R. Panaris, *J. Appl. Polym. Sci.*, **16**, 315, 1972.
321. S. Hattori, H. Endoh, H. Nakahara, T. Kamata and M. Hamashima, *Polymer J.*, **10**, 173, 1978.
322. M. J. Shepherd and J. Gilbert, *J. Chromatogr.*, **178**, 435, 1979.
323. J. Gilbert, M. J. Shepherd and M. A. Wallwork, *J. Chromatogr.*, **193**, 235, 1980.
324. R. Salovey and R. C. Gebauer, *J. Polym. Sci., Pt A1*, 1533, 1972.
325. S. Mori, *J. Chromatogr.*, **157**, 75, 1978.
326. J. Janca, S. Pokorny and M. Kolinsky, *J. Appl. Polym. Sci.*, **23**, 1811, 1979.
327. C. M. L. Atkinson and R. Dietz, *Europ. Polym. J.*, **15**, 21, 1979.
328. R. Dietz and M. A. Francis, *Polymer*, **20**, 450, 1979.
329. R. Dietz, *J. Appl. Polym. Sci.*, **25**, 951, 1980.
330. S. H. Agarwal and R. S. Porter, *J. Appl. Polym. Sci.*, **25**, 173, 1980.
331. K. J. Bombaugh, W. A. Dark and J. N. Little, *Anal. Chem.*, **41**, 1337, 1969.
332. C. M. L. Atkinson, R. Dietz and M. A. Francis, *Polymer*, **21**, 891, 1980.
333. A. Revillon, B. Dumont and A. Guyot, *J. Polym. Sci., Polym. Chem. Edn.*, **14**, 2263, 1976.
334. H. J. Holle and B. R. Lehnen, *Europ. Polym. J.*, **11**, 663, 1975.
335. K. Dodgson, D. Sympson and J. A. Semlyen, *Polymer*, **19**, 1285, 1978.
336. L. Mandik, A. Foksora and J. Foltyn, *J. Appl. Polym. Sci.*, **24**, 395, 1979.
337. I. A. Wachtina, A. P. Andrejew and O. G. Tarakanaw, *Plast. Kant.*, **26**, 626, 1979.

338. L. Segal, in *Advances in Chromatography*, Vol. 12, J. C. Giddings *et al.* (Eds.), Marcel Dekker, Inc., NY, 1975, p. 31.
339. K. E. Almin, K. E. Ericksson and B. A. Pettersson, *J. Appl. Polym. Sci.*, **16**, 2583, 1972.
340. Y. T. Bao, A. Bose, M. R. Ladisch and G. T. Tsoa, *J. Appl. Polym. Sci.*, **25**, 263, 1980.
341. D. Berek, G. Katuscakova and I. Novak, *Brit. Polym. J.*, **9**, 62, 1977.
342. C. Holt, W. Mackie and D. B. Seller, *Polymer*, **19**, 1421, 1978.
343. R. C. Rowe, *J. Pharm. Pharmacol.*, **32**, 116, 1980.
344. M. Papontonakis, *Tappi*, **63**, 65, 1980.
345. N. Morohoshi and W. G. Glasser, *Wood Sci. Technol.*, **13**, 249, 1979.
346. J. J. Minor, *J. Liquid Chromatogr.*, **2**, 309, 1979.
347. C. Y. Cha, *Polym. Lett.*, **7**, 343, 1969.
348. A. Domard, M. Rinaudo and C. Rochas, *J. Polym. Sci., Polym. Phys. Edn.*, **17**, 673, 1979.
349. N. Onda, K. Furusawa, N. Yamaguchi and S. Komuro, *J. Appl. Polym. Sci.*, **23**, 3631, 1979.
350. P. P. Nefedor, M. A. Lazareva, B. G. Belenkii, S. Y. Frenkel and M. M. Koton, *J. Chromatogr.*, **170**, 11, 1979.
351. H. J. Mencer and Z. G. Gallot, *J. Liquid Chromatogr.*, **2**, 649, 1979.
352. P. L. Dubin, S. K. Koontz and K. L. Wright, *J. Polym. Sci., Polym. Chem. Edn.*, **15**, 2047, 1977.
353. P. L. Dubin, *J. Liquid Chromatogr.*, **3**, 623, 1980.
354. W. J. Connors, S. Sarkanen and J. L. McCarthy, *Holzforschung*, **34**, 80, 1980.
355. P. P. Nefedor, M. A. Lazareva, B. G. Belenkii, S. Y. Frenkel and M. M. Koton, *J. Chromatogr.*, **170**, 11, 1979.
356. M. J. R. Canton, R. S. Porter and J. F. Johnson, *J. Polym. Sci.*, **C16**, 13, 1967.
357. C. Huber and K. H. Lederer, *J. Polym. Sci., Polym. Lett. Edn.*, **18**, 535, 1980.
358. B. M. E. Van Der Hoff and V. K. Chopra, *J. Appl. Polym. Sci.*, **23**, 2373, 1979.
359. L. Bartha, Gy. Vigh and N. Nemes, *Hung. J. Ind. Chem.*, **7**, 367, 1979.
360. R. L. Bartosiewiez, C. Booth and A. Marshall, *Europ. Polym. J.*, **10**, 783, 1974.
361. J. V. Dawkins, J. W. Maddock and A. Nevin, *Europ. Polym. J.*, **9**, 327, 1973.
362. J. Brzezinski and Z. Dobkowski, *Europ. Polym. J.*, **16**, 85, 1980.
363. M. J. Bowden and L. F. Thompson, *J. Appl. Polym. Sci.*, **19**, 905, 1975.
364. T. E. Attwood, T. King, I. D. McKenzie and J. B. Rose, *Polymer*, **18**, 365, 1977.
365. G. Christensen and P. Fink-Jensen, *J. Chromatogr. Sci.*, **12**, 59, 1974.
366. D. G. Lesnini, *J. Paint. Technol.*, **38**, 498, 1966.
367. L. Mandik, *Prog. Org. Coat.*, **5**, 131, 1977.
368. I. Ishigaki, Y. Morita, K. Nishimura and A. Ito, *J. Appl. Polym. Sci.*, **18**, 1927, 1974.
369. R. L. Miller, *Chromatogr. Newsletter*, **9**, 13, 1981.
370. R. A. Korus and R. W. Rosser, *Anal. Chem.*, **50**, 249, 1978.
371. J. A. P. P. VanDijk, W. C. M. Henkens and J. A. M. Smit, *J. Polym. Sci., Polym. Phys. Edn.*, **14**, 1485, 1976.
372. T. I. Obiaga and M. Wayman, *J. Appl. Polym. Sci.*, **18**, 1943, 1974.

373. *Sephadex® Gel Filtration Theory and Practice*, Pharmacia Fine Chemicals, Uppsala, Sweden, December 1977.
374. *Gel Filtration Theory and Practice*, Pharmacia Fine Chemicals, Uppsala, Sweden, 1975.
375. *A Laboratory Manual on Gel Chromatography*, Bio-Rad Laboratories, Richmond, California, US, 1975.
376. V. Tomasek, in *Laboratory Handbook of Chromatographic and Allied Methods*, O. Mikes (Ed.), Ellis Horwood Ltd, Chichester, UK, 1979, p. 334.
377. G. K. Ackers, in *Advances in Protein Chemistry*, Vol. 24, C. B. Vantinsen, J. T. Edsall and F. M. Richards (Eds.), Academic Press, NY, 1970, p. 343.
378. F. E. Regnier and K. M. Gooding, *Anal. Biochem.*, **103**, 1, 1980.
379. B. Stenlund, in *Advances in Chromatography*, Vol. 14, J. C. Giddings, E. Grushka, J. Cazes and P. R. Brown (Eds.), Marcel Dekker, Inc., NY, 1976, p. 37.
380. A. R. Cooper and D. P. Matzinger, *Amer. Lab.*, 13, January 1977.
381. A. R. Cooper and D. S. Van Derveer, *J. Liquid Chromatogr.*, **1**, 693, 1978.
382. K. G. Forss and B. G. Stelund, *J. Polym. Sci.*, **42**, 951, 1973.
383. P. A. Neddermeyer and L. B. Rogers, *Anal. Chem.*, **40**, 755, 1968.
384. M. Rinaudo and J. Desbrieres, *Europ. Polym. J.*, **16**, 849, 1980.
385. T. Lindstrom, A. DeRuvo and C. Soremark, *J. Polym. Sci., Polym. Chem. Edn.*, **15**, 2029, 1977.
386. A. W. Wolkoff and R. M. Larose, *J. Chromatogr. Sci.*, **14**, 51, 1976.
387. D. E. Brown and R. E. Lowry, *J. Polym. Sci., Polym. Chem. Edn.*, **17**, 1039, 1979.
388. A. H. Abdel-Alim and A. E. Hamielec, *J. Appl. Polym. Sci.*, **18**, 297, 1974.
389. A. L. Spatorico and G. L. Beyer, *J. Appl. Polym. Sci.*, **19**, 2933, 1975.
390. A. R. Cooper and D. P. Matzinger, *J. Appl. Polym. Sci.*, **23**, 419, 1979.
391. H. G. Barth, *J. Chromatogr. Sci.*, **18**, 409, 1980.
392. R. P. W. Scott and P. Kucera, *J. Chromatogr. Sci.*, **125**, 251, 1976.
393. F. A. Buytenhuys and F. P. B. Van der Maeden, *J. Chromatogr. Sci.*, **149**, 489, 1978.
394. C. P. Talley and L. M. Bowman, *Anal. Chem.*, **51**, 2239, 1979.
395. M. J. Telepchak, *J. Chromatogr.*, **83**, 125, 1973.
396. A. R. Cooper and D. P. Matzinger, in *Chromatography of Synthetic and Biological Polymers*, Vol. 1, R. Epton (Ed.), Ellis Hardwood Ltd, Chichester, UK, 1978, p. 350.
397. C. W. Hiatt, A. Shelokov, E. J. Rosenthal and J. M. Galimore, *J. Chromatogr.*, **56**, 362, 1971.
398. G. L. Hawk, J. A. Cameron and L. B. Dufault, *Prep. Biochem.*, **2**, 193, 1972.
399. H. D. Crone, R. M. Dawson and E. M. Smith, *J. Chromatogr. Sci.*, **103**, 71, 1975.
400. T. Mizutani and A. Mizutani, *J. Chromatogr. Sci.*, **111**, 214, 1975.
401. R. C. Collins and W. Haller, *Anal. Biochem.*, **54**, 47, 1973.
402. M. J. Frenkel and R. J. Blagrove, *J. Chromatogr.*, **111**, 397, 1975.
403. C. Persiani, P. Cukor and K. French, *J. Chromatogr. Sci.*, **14**, 417, 1976.
404. R. J. Blagrove and M. J. Frenkel, *J. Chromatogr.*, **132**, 399, 1977.
405. A. C. M. Wu, W. A. Bough, E. C. Conrad and K. E. Alden, Jr., *J. Chromatogr.*, **128**, 87, 1976.

406. S. H. Chang, K. M. Gooding and F. E. Regnier, *J. Chromatogr.*, **125**, 103, 1976.
407. D. L. Gooding, C. Chatfield and B. Coffin, *Amer. Lab.*, p. 48, August 1980.
408. H. G. Barth and F. E. Regnier, *J. Chromatogr.*, **192**, 275, 1980.
409. C. D. Chow and G. L. Jewett, *J. Liquid Chromatogr.*, **3**, 419, 1980.
410. H. Englehardt and D. Mathes, *J. Chromatogr.*, **142**, 311, 1977.
411. P. Roumeliotis and K. K. Unger, *J. Chromatogr.*, **185**, 445, 1979.
412. D. E. Schmidt Jr., R. W. Glese, D. Conron and B. L. Karger, *Anal. Chem.*, **52**, 177, 1980.
413. T. Hashimoto, H. Sasaki, M. Aiura and Y. Kato, *J. Polym. Sci., Polym. Phys. Edn.*, **16**, 1789, 1978.
414. K. Fukano, K. Komiya, H. Sasaki and T. Hashimoto, *J. Chromatogr.*, **166**, 47, 1978.
415. Y. Kato, K. Komiya, H. Sasaki and T. Hashimoto, *J. Chromatogr.*, **190**, 297, 1980.
416. Y. Kato, K. Komiya, H. Sasaki and T. Hashimoto, *J. Chromatogr.*, **193**, 311, 1980.
417. Y. Kato, H. Sasaki, M. Aiura and T. Hashimoto, *J. Chromatogr.*, **153**, 546, 1978.
418. T. Hashimoto, H. Sasaki, M. Aiura and Y. Kato, *J. Chromatogr.*, **160**, 301, 1978.
419. Y. Kato, K. Komiya, H. Sasaki and T. Hashimoto, *J. Chromatogr.*, **193**, 458, 1980.
420. T. Imamaura, K. Konishi, M. Yokoyama and K. Konishi, *J. Biochem.*, **86**, 639, 1979.
421. Y. Kato, K. Komiya, H. Sasaki and T. Hashimoto, *J. Chromatogr.*, **193**, 29, 1980.
422. C. T. Wehr and S. R. Abbot, *J. Chromatogr.*, **185**, 453, 1979.
423. C. M. L. Atkinson, R. Dietz and M. A. Francis, *Polymer*, **21**, 891, 1980.
424. S. Kumio and V. Tsutsumi, *Toyo Soda Kenkyu Hokoku*, **24**(1), 43, 1980.
425. R. V. Vivilecchia, B. G. Lightbody, N. Z. Thimot and H. M. Quinn, *J. Chromatogr.*, **15**, 424, 1977.
426. C. Thrall and T. C. Spelsberg, in *Biological/Biomedical Applications of Liquid Chromatography*, G. L. Hawk (Ed.), Marcel Dekker, Inc., NY, 1979, p. 283.
427. B. S. Welinder, *J. Liquid Chromatogr.*, **3**, 1399, 1980.
428. R. Epton, C. Holloway and J. V. McLaren, *J. Appl. Polym. Sci.*, **18**, 179, 1974.
429. R. Epton, C. Holloway and J. V. McLaren, *J. Chromatogr.*, **90**, 249, 1974.
430. R. Epton, *Aldrichimica Acta*, **9**, 23, 1976.
431. J. Coupek, M. Krivakova and S. Pokorny, *J. Polym. Sci., Symp. No. 42*, 185, 1973.
432. Lachema, Brno, Czechoslovakia.
433. Showa Denko KK, Tokyo, Japan.
434. F. E. Regnier and K. M. Gooding, *Anal. Biochem.*, **103**, 1, 1980.
435. J. L. DiCesare, *Chromatography Newsletter*, **8**, 52, 1980.
436. W. Heitz and P. Die, *Die Makromol. Chemie.*, **176**, 657, 1975.
437. C. D. Chow, *J. Chromatogr.*, **114**, 486, 1975.
438. I. H. Coopes, *J. Polym. Sci., Symp. No. 49*, 97, 1975.
439. I. Axelsson, *J. Chromatogr.*, **152**, 21, 1978.

440. J. Brower, *J. Chromatogr.*, **179**, 342, 1979.
441. L. Hazel, *J. Chromatogr.*, **160**, 59, 1978.
442. L. A. Huff and R. L. Easterday, *J. Liquid Chromatogr.*, **1**, 811, 1978.
443. L. A. Huff, *Prep. Biochem.*, **8**, 99, 1978.
444. R. Miyatake and J. Kumanotani, *Die Makromol. Chemie.*, **177**, 2749, 1976.
445. J. M. Curling, J. Berglof, L. O. Linquist and S. Eriksson, *Vox. Sang.*, **33**, 97, 1977.
446. R. Casey and M. W. Rees, *Anal. Biochem.*, **95**, 397, 1979.
447. A. R. Cooper and D. S. Van Derveer, in *Size Exclusion Chromatography (GPC)*, T. Provder (Ed.), *ACS Symposium Series No. 138*, American Chemical Society, Washington, DC, 1980, p. 297.
448. G. H. Fleet and D. J. Manners, *Biochem. Soc. Trans.*, **3**, 982, 1975.
449. N. K. Sabbagh and I. S. Fagerson, *J. Chromatogr.*, **120**, 55, 1976.
450. W. Brown and K. Chitumbo, *Chemica Scripta*, **2**, 88, 1972.
451. J. Cazes and D. R. Gaskill, *Separ. Sci.*, **2**, 421, 1967.
452. W. B. Smith and A. Kollmansberger, *J. Phys. Chem.*, **67**, 4157, 1965.
453. J. Cazes and D. R. Gaskill, *Separ. Sci.*, **4**, 15, 1969.
454. A. Lambert, *J. Appl. Polym. Sci.*, **20**, 305, 1970.
455. A. Lambert, *Anal. Chim. Acta.*, **53**, 63, 1971.
456. J. G. Hendrikson and J. C. Moore, *J. Polym. Sci.*, *Pt A1*,**4**, 167, 1966.
457. J. G. Hendrickson, *Anal. Chem.*, **40**, 49, 1968.
458. G. D. Edwards and Q. Y. Ng, *J. Polym. Sci.*, *Pt C*, **21**, 105, 1968.
459. S. Mori and A. Yamakara, *J. Liquid Chromatogr.*, **3**, 329, 1980.
460. J. C. Giddings, E. Kucera, L. P. Russell and M. N. Myers, *J. Phys. Chem.*, **72**, 4397, 1968.
461. S. Hjerten, *J. Chromatogr.*, **50**, 189, 1970.
462. S. Mori, *J. Chromatogr.*, **192**, 295, 1980.
463. E. Ozaki, K. Saitoh and N. Suzuki, *J. Chromatogr.*, **177**, 122, 1979.
464. J. E. Figueruelo, V. Soria and A. Campos, *J. Liquid Chromatogr.*, **3**, 367, 1980.
465. R. E. Majors and E. L. Johnson, *J. Chromatogr.*, **167**, 17, 1978.
466. G. Dallas and S. D. Abbott, on *Liquid Chromatographic Analysis of Food and Beverages*, Vol. 2, Academic Press, NY, 1979, p. 509.
467. R. A. Shoemaker, *J. Appl. Polym. Sci.*, *Appl. Polym. Symp. 34*, 139, 1978.
468. Shodex Technical Bulletin No. 110, May 1978.
469. Y. Kato, S. Kido, H. Watanabe, M. Yamamoto and T. Hashimoto, *J. Appl. Polym. Sci.*, **19**, 629, 1975.
470. S. A. Taleb-Bendiab and J. M. Vergnand, *J. Appl. Polym. Sci.*, **25**, 499, 1980.
471. A. Bledzki and M. I. Prusinska, *J. Polym. Sci.*, *Polym. Chem. Edn.*, **16**, 107, 1978.
472. L. E. Brydia and O. M. Garty, *J. Polym. Sci.*, *Polym. Chem. Edn.*, **18**, 1577, 1980.

INDEX

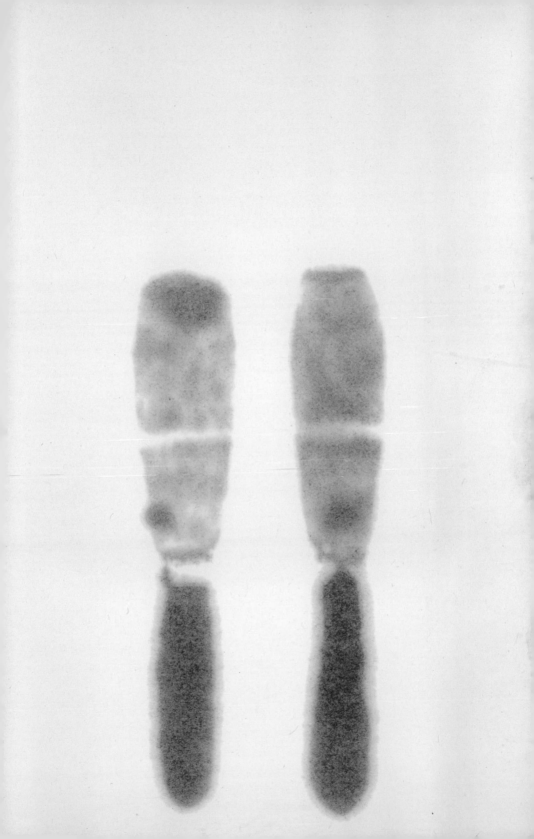

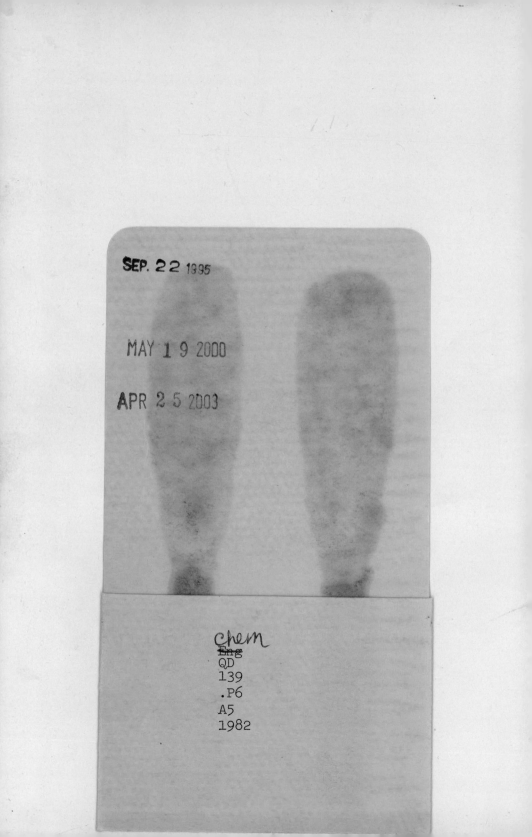